Einstein's Mistakes

in

Relativity

by

Karunesh M. Tripathi

CONTENTS

Preface i

Special Relativity

MISTAKES & OBJECTIONS

Chapter 1: Einstein's 1905 Paper Rides on a Kinematical Mistake to Get the Lorentz Transformation 1

Chapter 2: Einstein's 1905 Paper Rides on Incorrect Angular Frequency to Forecast Redshift 14

Chapter 3: $E = mc^2$ Used Indiscriminately Despite Its Derivations by Einstein Being Incorrect and Rudimentary 21

Chapter 4: Einstein's Incorrect Derivation of the Lorentz Transformation in his 1916 Book 29

Chapter 5: Spacetime in Question for Permitting Infinite Transformations for an event 40

INCORRECT APPLICATIONS
(OF INVALID LORENTZ TRANSFORMATION)

Chapter 6: The Unsound Structure of Relativity 51

Chapter 7: Manoeuvring of Results in Relativity by Manipulating Inputs 59

Chapter 8: How Relations for Light Forced upon Other Bodies Goes Unnoticed in Relativity 69

Chapter 9: Misconceptions about Time Dilation and Length Contraction 73

Chapter 10: Commensurate Validations Required for Theory of Relativity 80

Chapter 11: Part Application of Relativity Lead to Errors & Paradox in Doppler Effect and Incorrect Explanation for Muons Reaching Earth 88

Chapter 12: How Einstein Interchanged the Object and Result of Transformation to Show Length Contraction and Time Dilation in Moving Frames 99

Chapter 13: Twin Paradox and Travelling into Future, Perpetrated by Einstein, are Misinterpretations 105

Chapter 14: Relativity Can Neither Extend Muon's Lifespan Nor Shorten its Path. 108

SUGGESTIONS

Chapter 15: The Set of Postulates is Required to be Expanded 114

Chapter 16: Lorentz Transformation with Oblique Observation of a Moving Light Signal 117

Chapter 17: Maxwell's Equation for EM Waves—Does It Hold Good for Obliquely Moving Observers? 128

Chapter 18: Correction of a Flaw in Relativity Eliminates Wigner-Thomas Rotation 135

General Relativity

Chapter 19: Mistakes of and Questions on General Relativity 150

PREFACE

Einstein contributed to physics enormously by his work on photo electricity which also proved to be a solid stepping stone for quantum physics. His work on Relativity, however, emerges out to be more of a misleading force on physics that prevents the modern domains – the quantum physics, particle physics, astrophysics and cosmology etc. - from sailing in the right direction.

Serious mistakes – technical as well as conceptual - have been committed by Einstein in his works on Relativity, starting from his very first and the famous 1905 paper. The mistakes and the conceptual gaps are so obvious that their blind following by physicists from generations to generations is not only surprising but also frustrating.

It is clarified that the mistakes in derivation of the Lorentz Transformation for light in section 3 of his 1905 paper do not undermine the validity of the transformation, as these have successfully been derived by other methods, including the one presented in chapter 1 of this book.

When one applies the transformation to other-than-light events, for which the Lorentz Transformation have never been and cannot be derived , inconsistencies arise in the accepted notions of 'length contraction and time dilation due to motion'. However, Einstein himself indulged in such a departure in the very next section (4) of his 1905 paper, applying the Lorentz Transformation to rigid sphere, to show length contraction. Then after, he did it innumerable number of times, and unfortunately, the mistakes continue to be committed from generations to generations.

Given the image Einstein carries and the universal acceptance of his incorrect steps as correct, it is extremely important to bring these mess ups to notice of all so that the younger generations do not grow up with incorrect mathematics and physics, as has so far been happening.

Some of them are captured in the book which is a compilation of many articles I wrote over time. Being it so, repetition of the core ideas and some parts of text has been unavoidable. However, reader would find them fitting well in the context of individual chapters.

Further, the second postulate of theory of Special is too general to discreetly define the theory with its limitations, leading to misconceptions and paradoxes. Chapter 15 discusses and brings out the corrections/augmentations required in the postulates to remove the inconsistencies.

The mistakes of and questions on the theory of General Relativity have also been discussed in chapter 19.

Special Relativity

MISTAKES & OBJECTIONS

CHAPTER 1

Einstein's 1905 Paper Rides on a Kinematical Mistake to Get the Lorentz Transformation

Abstract:

Einstein's derivation of the Lorentz transformation in section 3 of his famous 1905 paper is based on kinematics of light rays in X, Y and Z directions. However, the velocity of light worked out in Y and Z directions, with respect to the frame moving in X direction, and as observed from the stationary frame, is found to be incorrect. This renders the entire derivation invalid.

It is further pointed out that even if the mistake was ignored, the relations are achieved by taking a time that is for travel of light from origin to a specific location (mirror). This limits the applicability of the relations only to events of light. Despite the restriction, however, Einstein extensively applied the relations to rigid bodies and moving clocks, even in the same paper.

It is also discussed as to how, in the derivation of his 1916 book, he used the relations derived for a moving light signal, on rigid bodies.

Further, a new objection-free method has been presented based on kinematics, using the principle of reciprocity of relative velocity between the two frames.

Introduction:

Einstein, in his 1905 paper, presented the theory of Special Relativity based on the principle that the time of travel of light between two locations was not the same in both the directions (of travel), except when the locations were in the observer's stationary frame. The corresponding exercise, appearing in section 3 of his paper (discussed below), culminated into the Lorentz transformation.

The derivation is all about equating, in a moving frame, the times of travel of light between two points, in the two directions, after casting them as functions of their own corresponding values, along with the distance between the two locations, as observed from the stationary frame. However, a kinematical mistake is found in calculation of the relative velocity of light in Y and Z directions, as observed from the stationary frame, and it renders the derivation invalid.

Further, even if the mistake is ignored for a moment, the Lorentz transformation relations have been achieved by taking a time that is for travel of light to a specific location. This limits the applicability of the relations only to events of light.

Working towards the solution, a new objection-free method has been presented towards the end, based on the reciprocity of relative velocity between the two frames, by kinematics.

The Section 3 of the paper is reproduced below in italics. My observations are intermittently placed, in normal font, between two dotted lines, as the matter progresses.

$3. Theory of the Transformation of Co-ordinates and Times from a Stationary System to another System in Uniform Motion of Translation Relatively to the Former

Let us in "stationary" space take two systems of co-ordinates, i.e. two sys- tems, each of three rigid material lines, perpendicular to one another, and issuing from a point. Let the axes of X of the two systems coincide, and their axes of Y and Z respectively be parallel. Let each system be provided with a rigid measuring-rod and a number of clocks, and let the two measuring-rods, and likewise all the clocks of the two systems, be in all respects alike.

Now to the origin of one of the two systems (k) let a constant velocity v be imparted in the direction of the increasing x of the other stationary system (K), and let this velocity be communicated to the axes of the co-ordinates, the relevant measuring-rod, and the clocks. To any time of the stationary system K there then will correspond a definite position of the axes of the moving system, and from reasons of symmetry we are entitled to assume that the motion of k may be such that the axes of the moving system are at the time t (this "t" always denotes a time of the stationary system) parallel to the axes of the stationary system.

We now imagine space to be measured from the stationary system K by means of the stationary measuring-rod, and also from the moving system k by means of the measuring-rod moving with it; and that we thus obtain the co-ordinates x, y, z, and ξ, η, ζ respectively. Further, let the time t of the stationary system be determined for all points thereof at which there are clocks by means of light signals in the manner indicated in §1; similarly let the time τ of the moving system be determined for all points of the moving system at which there are clocks at rest relatively to that system by applying the method, given in §1, of light signals between the points at which the latter clocks are located.

To any system of values x, y, z, t, which completely defines the place and time of an event in the stationary system, there belongs a system of values ξ, η, ζ, τ, determining that event relatively to the system k, and our task is now to find the system of equations connecting these quantities.

In the first place it is clear that the equations must be linear on account of the properties of homogeneity which we attribute to space and time.

If we place $x' = x - vt$ it is clear that a point at rest in the system k must have a system of values x', y, z, independent of time. We first define τ as a function of x', y, z, and t. To do this we have to express in equations that τ is nothing else than the summary of the data of clocks at rest in system k, which have been synchronized according to the rule given in §1.

From the origin of system k let a ray be emitted at the time τ_0 along the X-axis to x', and at the time τ_1 be reflected thence to the origin of the co-ordinates, arriving there at the time τ_2; we then must have $\frac{1}{2}(\tau_0 + \tau_2) = \tau_1$, or, by inserting the arguments of the function τ and applying the principle of the constancy of the velocity of light in the stationary system:—

$$\frac{1}{2}\left[\tau(0,0,0,t) + \tau\left(0,0,0, t + \frac{x'}{c-v} + \frac{x'}{c+v}\right)\right] = \tau\left(x', 0, 0, t + \frac{x'}{c-v}\right)$$

Hence, if x' be chosen infinitesimally small,

$$\frac{1}{2}\left(\frac{1}{c-v} + \frac{1}{c+v}\right)\frac{\partial \tau}{\partial t} = \frac{\partial \tau}{\partial x'} + \frac{1}{c-v}\frac{\partial \tau}{\partial t}$$

Or,

$$\frac{\partial \tau}{\partial x'} + \frac{v}{c^2 - v^2}\frac{\partial \tau}{\partial t} = 0$$

It is to be noted that instead of the origin of the co-ordinates we might have chosen any other point for the point of origin of the ray, and the equation just obtained is therefore valid for all values of x', y, z.
An analogous consideration—applied to the axes of Y and Z—it being borne in mind that light is always propagated along these axes, when viewed from the stationary system, with the velocity $\sqrt{c^2 - v^2}$ gives us

$$\frac{\partial \tau}{\partial y} = 0 \; , \frac{\partial \tau}{\partial z} = 0$$

Since τ is a *linear* function, it follows from these equations that

$$\tau = a\left(t - \frac{v}{c^2 - v^2}x'\right)$$

where a is a function $\varphi(v)$ at present unknown, and where for brevity it is assumed that at the origin of k, $\tau = 0$, when t = 0.

Observations 1:

i) The relation obtained just above is a general relation for any value of t and x', provided that at the origin of k, $\tau = 0$, when t = 0.
To check whether the requirement mentioned at the start is met with, let us work out the values of τ_0, τ_1 and τ_2. On substituting the parameters of τ, at the three instants, from the starting equation, we get the values as follows.

$$\tau_0 = a\left(t - \frac{v}{c^2 - v^2} \times 0\right) = at$$

$$\tau_1 = a\left(\left(t + \frac{x'}{c-v}\right) - \frac{v}{c^2 - v^2}x'\right) = a\left(t + \frac{c}{c^2 - v^2}x'\right)$$

$$\tau_2 = a\left(\left(t + \frac{x'}{c-v} + \frac{x'}{c+v}\right) - \frac{v}{c^2-v^2} \times 0\right) = a\left(t + \frac{2c}{c^2-v^2}x'\right)$$

Therefore,

$$\tau_0 + \tau_2 = at + a\left(t + \frac{2c}{c^2-v^2}x'\right) - 2a\left(t + \frac{c}{c^2-v^2}x'\right) = 2\tau_1$$

Thus the relation meets the stipulated requirement.

It is pointed out here that though the relation meets the assumptions it is derived from, it fails to get the Lorentz transformation, when correct values are inputted to it, as shown in my Observations 3 and 4 ahead. The reason is a mistake which is discussed at (ii) below.

ii) The statement that the velocity of light in Y and Z directions, when viewed from the stationary system, would always be $\sqrt{c^2 - v^2}$ is incorrect, as explained below.

First of all, it is clarified that at this stage, the value of relative velocity of light in Y or Z direction is to be taken **by kinematics and not relativity,** which is also obvious from the terms of time traversed by light moving in X direction, in the starting equation itself, such as $(t + \frac{x'}{c-v} + \frac{x'}{c+v})$ and $(t + \frac{x'}{c-v})$. Thus the relative velocity of light with respect to the moving frame can be both, less as well as more than c by kinematics.

Following the same principles, the relative velocity of light with respect to the origin of frame moving in X direction, as observed from the stationary frame, would always be $\sqrt{c^2 + v^2}$ (by kinematics) in the diagonal direction in XY plane or XZ plane respectively, as the two velocities, i.e. c and v, are mutually perpendicular to each other. However, the same velocity $\sqrt{c^2 + v^2}$ (in the diagonal direction) would appear as velocity c in the Y and Z directions (as the case may be), and as velocity v in the X direction, as its components.

To make it clearer, if we consider c and v as vectors, the relative velocity of light with respect to that of the moving frame, as seen from the stationary frame, is $(\vec{c} - \vec{v})$. When the two are collinear, as along X direction, these are simply added $(c + v)$ or subtracted

$(c - v)$, as also done by the author at start itself. On the other hand, when the two velocities are perpendicular to each other, with the light ray moving either in the Y direction or in the Z direction, the magnitude of the aforementioned relative velocity becomes $\sqrt{c^2 + v^2}$, irrespective of whether $\vec{v}$ is in (+)ve direction or in (-)ve direction. If, however, the relative velocity was sought only in the Y direction or only in the Z direction, it would always be c, as there is no component of velocity v in these directions.

Even if one argues that Einstein took the velocity of light in the diagonal direction as c relativistically (though it is not to be done at this stage), the velocity of light in in Y or Z direction would still continue to be c by the same principle of relativity. Thus the figure of $\sqrt{c^2 - v^2}$ worked out by Einstein finds no place, either relativistically or non-relativistically.

Although the correction does not alter the relations mentioned ahead i.e. $\frac{\partial \tau}{\partial y} = 0$, $\frac{\partial \tau}{\partial z} = 0$ yet the derivation ahead fails to get Lorentz transformation, as will be seen shortly.

With the help of this result we easily determine the quantities ξ, η, ζ by expressing in equations that light (as required by the principle of the constancy of the velocity of light, in combination with the principle of relativity) is also propagated with velocity c when measured in the moving system. For a ray of light emitted at the time $\tau = 0$ in the direction of the increasing ξ

$$\xi = c\tau \text{ or } \xi = ac\left(t - \frac{v}{c^2 - v^2}x'\right)$$

But the ray moves relatively to the initial point of k, when measured in the stationary system, with the velocity $c - v$, so that

$$\frac{x'}{c - v} = t$$

If we insert this value of t in the equation for ξ, we obtain

$$\xi = a\frac{c^2}{c^2 - v^2}x'$$

Observations 2:

As already mentioned in Observations 1 (i) above, the expression of τ, as a function of t and x', is of a general nature, given the theory of transformation proposed. However, selecting t as the time for light to travel a distance of x' curtails the applicability of the resultant relation only to events of light.

Despite such a limiting condition, Einstein used the relations extensively on all space–time sets, the first being on a rigid sphere in the next section 4 titled "Physical Meaning of the Equations Obtained in Respect to Moving Rigid Bodies and Moving Clocks". The deviation is also practiced universally.

The above issue (of deviation from assumptions of derivation), however, gets relegated behind, when one discovers a mistake, which is already explained in my Observations 1 (ii) above.

In an analogous manner we find, by considering rays moving along the two other axes, that

$$\eta = c\tau = ac\left(t - \frac{v}{c^2 - v^2}x'\right)$$

when

$$\frac{y}{\sqrt{c^2-v^2}} = t, \quad x' = 0$$

Thus

$$\eta = a\frac{c}{\sqrt{c^2-v^2}}y \text{ and } \zeta = a\frac{c}{\sqrt{c^2-v^2}}z$$

Substituting for x' its value, we obtain

$$\tau = \emptyset(v)\beta(t - vx/c^2),$$

$$\xi = \emptyset(v)\beta(x - vt),$$

$$\eta = \emptyset(v)y,$$

$$\zeta = \emptyset(v)z,$$

where

$$\beta = \frac{1}{\sqrt{1 - v^2/c^2}}$$

Observations 3:

It has been explained in my Observations 1 (ii) above that the velocity of light in Y and Z directions is incorrectly taken as $\sqrt{c^2 - v^2}$, and it should correctly be c from simple kinematics. On applying this correction by replacing $\sqrt{c^2 - v^2}$ with c, the expressions of η and ζ become as follows.

$$\eta = a\frac{c}{c}y = ay \text{ and } \zeta = a\frac{c}{c}z = az$$

Further, in the above relations, the function $\emptyset(v)$ has been taken as follows.

$$\emptyset(v) = a\frac{c}{\sqrt{c^2 - v^2}} = a\beta$$

When the above correction is applied, one gets $\emptyset(v) = a\frac{c}{c} = a$.

Therefore, to correct the above-mentioned 4 relations, one has to multiply the RHS by β. On doing so, one gets,

$$\tau = \emptyset(v)\beta^2(t - vx/c^2),$$

$$\xi = \emptyset(v)\beta^2(x - vt),$$

$$\eta = \emptyset(v)\beta y,$$

$$\zeta = \emptyset(v)\beta z,$$

Notes:

The derivation goes on further to find out the value of $\emptyset(v)$ which turns out to be 1.

Observations 4:

When the calculated value of $\emptyset(v)$ as 1 is substituted, the relations become as follows.

$$\tau = \beta^2(t - vx/c^2),$$

$$\xi = \beta^2(x - vt),$$

$$\eta = \beta y,$$

$$\zeta = \beta z,$$

These are different from the Lorentz transformation relations and also cannot be the targeted relations, as these would not conform to constancy of light speed c in the moving frame.

Thus the assumptions made for transformation of coordinates and times fail to get the Lorentz transformation.

Does it mean that the Lorentz Transformation are Incorrect?

No. The above exercise is only a failed attempt to derive the Lorentz transformation for light by kinematics, which has otherwise been established by electrodynamics. However, these relations for events other than those of light are not yet derived, though used universally. The same Lorentz transformation relations can be derived for such events too, but with a different postulate/assumption.

Alternative Methods:

1. Einstein's 1916 Book:

Annexure I of the book also presents a derivation by kinematics. However, the derivation starts with building up equations (1) to (5) for a moving light signal, but soon thereafter digresses to apply the set of equations (5) on other-than-light objects such as origin of the moving frame, meter-rods and clocks placed in the stationary and moving frames etc.

Thus, this derivation too is not correct, and therefore, is unsuitable for adoption.

2. A New Derivation Based on Reciprocity of Velocity:

Two of Einstein's derivations of the Lorentz transformation relations by kinematics are found to be incorrect, as explained above. On the other hand, by application of the principle of velocity reciprocity between the two frames, it is possible to achieve the relations by kinematics.

The same is presented below [2].

The following fig.1 may be referred.

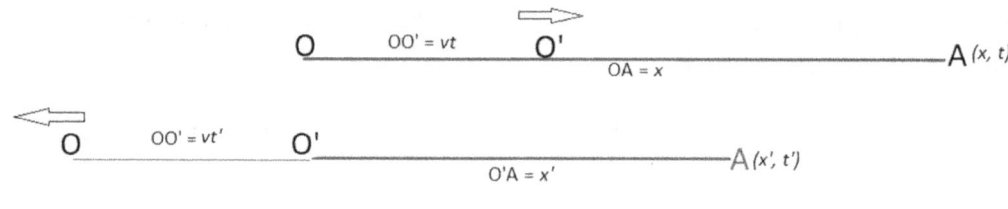

Fig.1

The point A is the event which may be of a moving light signal or any arbitrary distance-time set. So, it is shown merely as (x, t) in the non-primed frame and as (x', t') in the primed frame. In case of light, the two parameters of the event are related as $(x = ct)$ and $(x' = ct')$ respectively.

The points O and O' are the locations of origins of the non-primed and the primed frames respectively, which are in relative motion with respect to each other with a uniform relative velocity v in the direction of location of the event A from origin O i.e. v is along the line OO'A. At time $t = t' = 0$, both the origins were coincident, and at the same time, one of the frames starts its motion. Either of O and O' could be taken as moving toward A, along the line OO'A. In the instant case, however, O' is considered to be so.

Both the diagrams show the locations of O, O' and A at a particular time t after the start.

The top one represents O as stationary and O' as moving, with the observer at O watching O' and A.

Similarly, the bottom one represents O' as stationary and O as moving in the opposite direction, with the observer at O' watching O and A.

x is the distance of the event A at time t in the non-primed frame, and similarly, x' is the corresponding distance of the event A at time t' in the primed frame.

Both the representations are equally correct and interchangeable, by the essence of relativity i.e. reciprocity.

Applying classical kinematics, one may write the following relations for the two cases respectively.

$$\left.\begin{array}{l} O'A = x' = x - vt \\ OA = x = x' + vt' \end{array}\right\}$$

However, according to the Special Relativity, the distance and time in the moving frame are so modified that these conform to the postulate of constancy of light speed (in vacuum) in all inertial frames, and/or to the Lorentz Transformation Condition.

Therefore, a bridging parameter, in the form of a constant, need to be applied to both the relations to strike conformity.

Let a be such a constant to be applied to both, as follows.

$$\left.\begin{array}{l} x' = a(x - vt) \\ x = a(x' + vt') \end{array}\right\} \quad \ldots \ldots (1)$$

On separating t' and t, one-by-one, from the above two relations (1), one gets the following.

$$\left.\begin{array}{l} t' = a\left[t - \left(1 - \frac{1}{a^2}\right)\frac{x}{v}\right] \\ t = a\left[t' + \left(1 - \frac{1}{a^2}\right)\frac{x'}{v}\right] \end{array}\right\} \quad \ldots \ldots (2)$$

Since both the frames are inertial, the postulate stipulates that $x = ct$ as well as $x' = ct'$.

Using the above relations (1) and (2), in conjunction with the postulate, separately in the two frames i.e. $x = ct$ and $x' = ct'$, let us proceed as follows, on two separate threads (columns of the following table).

$x' = ct'$	$x = ct$
Substitute the expressions of x' and t' from relations (1) and (2) above to get $$a(x - vt) = c\left(a\left[t - \left(1 - \frac{1}{a^2}\right)\frac{x}{v}\right]\right)$$	Substitute the expressions of x and t from relations (1) and (2) above to get $$a(x' + vt') = c\left(a\left[t' + \left(1 - \frac{1}{a^2}\right)\frac{x'}{v}\right]\right)$$
Substitute x with ct to get $$a(ct - vt) = c\left(a\left[t - \left(1 - \frac{1}{a^2}\right)\frac{ct}{v}\right]\right)$$	Substitute x' with ct' to get $$a(ct' + vt') = c\left(a\left[t' + \left(1 - \frac{1}{a^2}\right)\frac{ct'}{v}\right]\right)$$
Divide both the sides of eq. by ct to get $$a\left(1 - \frac{v}{c}\right) = a\left[1 - \left(1 - \frac{1}{a^2}\right)\frac{c}{v}\right]$$ Or, $$1 - \frac{v}{c} = 1 - \left(1 - \frac{1}{a^2}\right)\frac{c}{v}$$ Or, $$a = \frac{1}{\sqrt{1 - \frac{v^2}{c^2}}} = \gamma$$	Divide both the sides of eq. by ct' to get $$a\left(1 + \frac{v}{c}\right) = a\left[1 + \left(1 - \frac{1}{a^2}\right)\frac{c}{v}\right]$$ Or, $$1 + \frac{v}{c} = 1 + \left(1 - \frac{1}{a^2}\right)\frac{c}{v}$$ $$a = \frac{1}{\sqrt{1 - \frac{v^2}{c^2}}} = \gamma$$
Substitute the above value of a in relations (1) and (2), and replace a with γ to get $$x' = \gamma(x - vt) \\ t' = \gamma\left(t - \frac{vx}{c^2}\right) \\ x = \gamma(x' + vt') \\ t = \gamma\left(t' + \frac{vx'}{c^2}\right)$$	Substitute the above value of a in relations (1) and (2), and replace a with γ to get $$x' = \gamma(x - vt) \\ t' = \gamma\left(t - \frac{vx}{c^2}\right) \\ x = \gamma(x' + vt') \\ t = \gamma\left(t' + \frac{vx'}{c^2}\right)$$

Thus the Lorentz transformation is achieved for both the frames separately, by the new method.

Conclusion:
The article shows as to how Einstein, in his 1905 paper, derived the Lorentz transformation by kinematics, with a mistake which has remained unnoticed by the Physics community.

His next attempt in his 1916 book is also riddled with mistakes, which have been pointed out briefly in the article.

The mistakes, however, do not undermine the validity of the Lorentz transformation, which has been established beyond doubt for light, by electrodynamics.

The essence of relativity i.e. reciprocity of the relative velocity between the two frames has been shown to be strong enough to throw out the Lorentz transformation by kinematics, in both the frames individually.

References:
1. A. Einstein, "On the Electrodynamics of Moving Bodies", June 30, 1905, distributed by http://www.fourmilab.ch/
2. Author's Book "Refining Relativity Part 1 (The Special Theory)", 2020.

CHAPTER 2

Einstein's 1905 Paper Rides on Incorrect Angular Frequency to Forecast Redshift

Abstract:

The relativistic Doppler Effect is explained from the theory of Relativity by applying an incorrect factor (γ) for the time dilation to the time period of the electromagnetic waves arriving the Earth from celestial sources. This leads to a forecast of redshift. However, if the correct ratio of time transformation i.e. $\sqrt{\frac{1-v/c}{1+v/c}}$ is applied, one gets a blueshift because the frequency, being inverse of the time period, changes in the ratio of $\sqrt{\frac{1+v/c}{1-v/c}}$. Thus Relativity, when applied correctly, leads to results just reverse of what is professed today.

A question may, however, arise as to how to reconcile the above statement with the redshift worked out by Einstein in his 1905 Paper, Section 7 titled "Theory of Doppler's Principle and of Aberration". The answer is he committed a mistake in working out the angular frequency of light in the moving (observer's) frame.

Introduction:

The Lorentz transformation, for the case of the primed frame moving with a uniform velocity v in (+)ve x-direction with respect to the non-primed frame (stationary), is as follows [1][2].

$$x' = \gamma(x - vt)$$

$$t' = \gamma\left(t - \frac{vx}{c^2}\right)$$

where $\gamma = 1/\sqrt{1 - v^2/c^2}$

In case of electromagnetic waves, since $x = ct$, the above relations get reduced to the following.

$$x' = \sqrt{\frac{1 - v/c}{1 + v/c}}\, x$$

$$t' = \sqrt{\frac{1 - v/c}{1 + v/c}}\, t$$

The relations show that the transformed distance as well as time traversed by the EM wave, in the frame co-directionally moving (observer) with the wave, get lesser than those existing in the stationary frame (source).

Applying the same result, the time period of the EM wave too would get reduced in the moving frame. This leads to conclusion that the frequency, being inverse of the time period, would increase for the moving frame i.e. observer, meaning a blueshift.

The same results are obtained by considering the source as the moving frame and the observer/receiver as the stationary frame.

The current practice, however, disregards the above facts and instead chooses to incorrectly apply a time dilation by a factor γ to the time period of the wave emitted by source (considered stationary) to work out its value in the moving (observer's) frame. This leads to redshift, which is taken as an evidence of correctness of the method. However, the fact remains that if the source was considered moving instead of the observer (in line with Relativity), the results change enormously, leading to a paradox, and this questions the validity of the existing practice.

Now, a question arises as to how Einstein managed to get a redshift on application of Relativity in his 1905 Paper, Section 7 titled "Theory of Doppler's Principle and of Aberration" [1]. The answer is he committed a mistake in working out the angular frequency of light in the moving (observer's) frame.

The same is explained below.

Discussion:

First, let us recall the parameters used in the paper.

The (t, x, y, z) are the parameters of the light wave in the stationary frame (source), and the corresponding parameters in the moving (observer's) frame are $(\tau, \xi, \eta, \varsigma)$.

The angular frequency parameters ω and ω' are similarly for the stationary and the moving frames respectively.

Similarly, (l, m, n) are the direction cosines of the wave normal (direction of ray) along the x, y and z axes respectively, and their corresponding values in the moving frame are (l', m', n').

Now, coming directly to the spot of the mistake, the following two relations (in bold) have been stated, which are correct **except the relation between ω' and ω**.

[**Note:** Please note that β has been used here for the Lorentz Factor, in place of the current notation of γ]

$$\Phi = \omega\left\{t - \frac{1}{c}(lx + my + nz)\right\}$$

$$\Phi' = \omega'\left\{\tau - \frac{1}{c}(l'\xi + m'\eta + n'\varsigma)\right\}$$

where

$$\omega' = \omega\beta(1 - lv/c)$$

$$l' = \frac{l - v/c}{1 - lv/c}$$

$$m' = \frac{m}{\beta(1 - lv/c)}$$

$$n' = \frac{n}{\beta(1 - lv/c)}$$

where $\beta = 1/\sqrt{1 - v^2/c^2}$

Observations:

The given expression of ω' is incorrect, as brought out below.

The times t and τ, in the stationary frame and the moving frame respectively, are related by

$$\tau = \beta(1 - lv/c)t$$

The same relation has also been used in the expressions for l', m' and n' whose denominators are $c\tau$, or $\beta(1 - lv/c)ct$.

Now, let the time period of the light wave in the stationary frame K and the moving frame k be T and T' respectively. Therefore, $\omega = 2\pi/T$ and $\omega' = 2\pi/T'$.

The transformation of time would apply to all timespans, including the time period T. Therefore, similar to the transformation $\tau = \beta(1 - lv/c)t$, the following relation is also true.

$$T' = \beta(1 - lv/c)T$$

Therefore,

$$\omega' = \frac{2\pi}{T'} = \frac{2\pi}{\beta(1 - lv/c)T} = \frac{\omega}{\beta(1 - lv/c)}$$

Thus the correct ratio of the angular frequencies in the two frames i.e. ω'/ω is just the inverse of what is worked out by Einstein.

Implications:

As a result, when the corrections are applied to the frequency, the forecast of a redshift in the light arriving from receding stars and galaxies, as experienced by the observers on the Earth, turn into a blueshift. The same is further corroborated below.

Einstein in his paper has gone ahead to reduce the general result to a case where the observer was moving in the direction of light received (from a receding luminous body) i.e. for $l = 1$.

Since he took $\omega'/\omega = \beta(1 - lv/c)$, it reduced to the following relation.

$$\nu' = \nu \sqrt{\frac{1 - v/c}{1 + v/c}}$$

Where ν' and ν are the frequencies in the observer's (moving) frame and the light-emitting-body's frame (stationary) respectively. That meant a redshift for observers on the Earth.

However, since the correct ratio of ω'/ω is just the inverse of what has incorrectly been taken, the above relation turns into

$$\nu' = \nu \sqrt{\frac{1 + v/c}{1 - v/c}}$$

which means a blueshift.

Conclusion:

The above exercise establishes beyond doubt that if the Special Relativity was correctly applied to the relative motion between observers on the Earth and the light emitting celestial bodies, the results were always blueshift. However, by incorrect application of a time dilation factor of γ, a redshift is shown to occur. While maintaining the theory of Relativity, if the source was considered moving and one applied the factor γ to its time, the results are entirely different. This leads to a paradox, declaring that the application of the theory is faulty.

On the other hand, if Relativity was correctly applied to either of the two frames, the results were the same.

It may be recalled that Einstein did not apply only the Lorentz Factor to the time of the stationary frame, to get the time of the moving frame. Instead, he has used the correct relation for time transformation, fully in accordance with the theory and its product, the Lorentz transformation. However, he failed to take note of the inverse proportionality between the angular frequency and the time period of waves.

The above discussions brings two options to the fore i.e. either the theory of Relativity is incorrect, or the notion of the redshift being on account of the theory is incorrect. The chances of the former are negligible, in view of the established constancy of light speed in uniformly moving frames. Therefore, we have to look for factors other than those of Relativity to find the answer for the redshift, which is currently thought of being on account of the so called relativistic Doppler Effect.

References:

1. A. Einstein, "On the Electrodynamics of Moving Bodies", June 30, 1905, distributed by http://www.fourmilab.ch/

A. Einstein's book "Relativity: The Special and The General Theory" 1916

CHAPTER 3

$E = mc^2$ Used Indiscriminately Despite Its Derivations by Einstein Being Incorrect and Rudimentary

Introduction:

The relation $E = mc^2$ is considered as one of the triumphant colours of science and the general public treat it as an indelible signature of Einstein on the canvas of knowledge. The physicists have used the equivalence in almost all the domains of modern physics viz. atomic physics, nuclear physics, particle physics, quantum physics, astrophysics and cosmology etc.

A natural question arises whether the relation's derivation is so broad based as to cover the diverse types of particles/energies bound/released by diverse types of forces/energies under diverse conditions. To find the answer, there cannot be a better document than the derivation carried out by the father of the relation i.e. Einstein. He published two papers presenting the derivations – one in 1905 and the second in 1935 – with the titles as follows.

1. "Does the Inertia of a Body Depend upon its Energy Content?" (27-Sep-1905)
2. "Elementary Derivation of the Equivalence of Mass and Energy" 1935

Both are discussed as follows.

1. The Paper of 1905:

The contents of the above quoted paper [1] are reproduced below in italics. My observations have been placed at the end between two dotted lines in normal font.

Let a system of plane waves of light, referred to the system of co-ordinates (x, y, z), possess the energy l ; let the direction of the ray (the wave-normal) make an angle ϕ with the axis of x of the system. If we introduce a new system of co-ordinates (ξ, η, ζ) moving in uniform parallel translation with respect to the system (x, y, z), and having its origin of co-ordinates in motion along the axis of x with the velocity v, then this quantity of light measured in the system (ξ, η, ζ) possesses the energy

$$l^* = l \frac{1 - \frac{v}{c} \cos \varphi}{\sqrt{1 - v^2/c^2}}$$

[**Note:** *The principle of the constancy of the velocity of light is of course contained in Maxwell's equations.*]

where c denotes the velocity of light. We shall make use of this result in what follows.

Let there be a stationary body in the system (x, y, z), and let its energy—referred to the system (x, y, z) be E_0. Let the energy of the body relative to the system (ξ, η, ζ) moving as above with the velocity v, be H_0.

Let this body send out, in a direction making an angle ϕ with the axis of x, plane waves of light, of energy(½)L measured relatively to (x, y, z), and simultaneously an equal quantity of light in the opposite direction. Meanwhile the body remains at rest with respect to the system (x, y, z). The principle of energy must apply to this process, and in fact (by the principle of relativity) with respect to both systems of co-ordinates. If we call the energy of the body after the emission of light E_1 or H_1 respectively, measured relatively to the system (x, y, z) or (ξ, η, ζ) respectively, then by employing the relation given above we obtain

$$E_0 = E_1 + \frac{1}{2}L + \frac{1}{2}L,$$

$$H_0 = H_1 + \frac{1}{2}L\frac{1-\frac{v}{c}\cos\varphi}{\sqrt{1-v^2/c^2}} + \frac{1}{2}L\frac{1+\frac{v}{c}\cos\varphi}{\sqrt{1-v^2/c^2}}$$

$$= H_1 + \frac{L}{\sqrt{1-v^2/c^2}}$$

By subtraction we obtain from these equations

$$H_0 - E_0 - (H_1 - E_1) = L\left\{\frac{1}{\sqrt{1-v^2/c^2}} - 1\right\}$$

The two differences of the form $H - E$ occurring in this expression have simple physical significations. H and E are energy values of the same body referred to two systems of co-ordinates which are in motion relatively to each other, the body being at rest in one of the two systems (system (x, y, z)). Thus it is clear that the difference $H - E$ can differ from the kinetic energy K of the body, with respect to the other system (ξ, η, ς), only by an additive constant C, which

depends on the choice of the arbitrary additive constants of the energies H and E. Thus we may place

$$H_0 - E_0 = K_0 + C,$$
$$H_1 - E_1 = K_1 + C,$$

since C does not change during the emission of light. So we have

$$K_0 - K_1 = L\left\{\frac{1}{\sqrt{1-v^2/c^2}} - 1\right\}$$

The kinetic energy of the body with respect to (ξ, η, ς) diminishes as a result of the emission of light, and the amount of diminution is independent of the properties of the body. Moreover, the difference

$K_0 - K_1$, like the kinetic energy of the electron (§ 10), depends on the velocity.

Neglecting magnitudes of fourth and higher orders we may place

$$K_0 - K_1 = \frac{1}{2}\frac{L}{c^2}v^2$$

From this equation it directly follows that:—

If a body gives off the energy L in the form of radiation, its mass diminishes by L/c^2. The fact that the energy withdrawn from the body becomes energy of radiation evidently makes no difference, so that we are led to the more general conclusion that

The mass of a body is a measure of its energy-content; if the energy changes by L, the mass changes in the same sense by $L/9 \times 10^{20}$, the energy being measured in ergs, and the mass in grammes.

It is not impossible that with bodies whose energy-content is variable to a high degree (e.g. with radium salts) the theory may be successfully put to the test.

If the theory corresponds to the facts, radiation conveys inertia between the emitting and absorbing bodies.

Observations:

1. Instead of emission of two light waves of energy L/2 in opposite directions, if one takes only one light wave of double the energy i.e. L emitted at the given angle φ, the last expression changes to as follows.

$$K_0 - K_1 = L\left\{\frac{1 - \frac{v}{c}\cos\varphi}{\sqrt{1 - v^2/c^2}} - 1\right\} = L\left\{\frac{1}{\sqrt{1 - v^2/c^2}} - 1 - \frac{\frac{v}{c}\cos\varphi}{\sqrt{1 - v^2/c^2}}\right\}$$

$$= \frac{1}{2}\frac{L}{c^2}v^2 - L\frac{v}{c}\cos\varphi\left(1 + \frac{1}{2}\frac{v^2}{c^2}\right)$$

Thus the equivalence becomes a function of φ, and the formula $E = mc^2$ is able to hold only for $\varphi = \frac{\pi}{2}, \frac{3\pi}{2}$.

2. The change in the total energy L of the emitted lights occurs solely due to transformation from stationary frame to moving frame. However, the entire change has been attributed to diminution of mass of the emitting body. This can only be termed as a motivated assumption, as brought out in the following para.

What if there was no moving observer and the emitted light was observed in the stationary frame itself? Obviously, there will not be any transformation and therefore, no change would occur in the energy L. It means that the mass of the emitting body would remain intact. This would contradict the theory itself.

3. Even with such grave contradictions in the exercise, the author has dreamt at the end of the paper that "the theory may be successfully put to the test" for even the radioactive salts/elements. Carrying on the dream, we have extensively been using it in modern physics and also been making all out efforts to show conformity, rather than reviewing it in the contexts used.

The Message:

The derivation presented by Einstein is found to be too deficient for the mass-energy formula to represent all areas, staring from (sub) atomic to astronomical levels. The very structure of attributing the change in energy of two light waves, arising out of Lorentz transformation of space and time, to that of kinetic energy of the emitting body supposedly on account of diminution of its mass, makes it irrelevant on even a slight change in the setup of assumptions, leave aside the complexities involved at atomic and subatomic levels. The same equivalence is used in the stress-energy-momentum tensors applied in General Relativity, which in turn is used to model phenomena at the Universe/Cosmos level.

There is thus an urgent need for the physicists to review the equivalence formula in the contexts these are being used, rather than carrying on with such a deficient exercise.

2. The Paper of 1935:

By this time, the concept of spacetime had significantly been popularized due to its cardinal role in the General Relativity.

For this reason probably, Einstein chose to present another derivation based on it. As will be shown ahead, he had taken the acceptance of the mass-energy equivalence for granted and simply pushed down his interpretation, though incorrectly, with the terms emanating from the Lorentz invariant of spacetime.

The salient steps are given below with my observations intermittently placed between two dotted lines.

The paper starts with the *"fundamental invariant of the Lorentz transformation"* as follows.

$$ds^2 = dt^2 - dx^2 - dy^2 - dz^2$$

or

$$ds = dt(1 - u^2)^{1/2}$$

where

$$u^2 = \left(\frac{dx}{dt}\right)^2 + \left(\frac{dy}{dt}\right)^2 + \left(\frac{dz}{dt}\right)^2$$

Further, It mentions - If one divides the components of the contravariant vector (dt, dx, dy, dz) by ds, one obtains the vector

$$\left(\frac{1}{(1-u^2)^{1/2}}, \frac{u_1}{(1-u^2)^{1/2}}, \frac{u_2}{(1-u^2)^{1/2}}, \frac{u_3}{(1-u^2)^{1/2}}\right)$$

Observations 1:

It is pointed out here that the vector obtained just above has its elements only as numbers without any dimension/unit. This is because **dt** carries a scale of **1/c** which makes it a distance equal to **cdt**, and the space elements already carry a scale of **1**. Thus, **ds** also carries the same unit. Therefore, division by **ds** cancels out the units and makes the vector dimensionless. Similarly, the terms u_1, u_2, u_3 and **u** are also dimensionless, as these are only fractions (with respect to **c**).

It is clarified here that though some may name the above vector as a four-velocity vector in a different sense, there is no element of velocity (distance covered per unit time) in any of its elements, and these are mere dimensionless numbers.

Further, it has been argued that if the vector belonged to a material particle of mass **m**, "*we obtain a vector connected with the motion of the particle by multiplying by* **m** *the four-vector of velocity that we have just written....*"

The first term of the vector, after multiplication by **m** becomes $\dfrac{m}{(1-u^2)^{1/2}}$, which on expansion and neglecting the third power of velocity becomes

$$m + \frac{1}{2}mu^2$$

It has further been argued that the second term is kinetic energy of the particle. Since the entire term was of energy, it is natural to "*ascribe to the mass-point in a state of rest the rest-energy* **m** *(with the usual time unit,* mc^2 *)*".

Observations 2:

As explained in my previous observations, the vector has no dimensions and calling it a four-velocity vector, in a different sense, does not make it a velocity for the purpose of calculating kinetic energy. Thus calling $\frac{1}{2}mu^2$ as kinetic energy is a mistake. Further, the "*usual time unit*" c^2 has incorrectly been assigned to first term **1**, which is dimensionless, as the unit is already cancelled on division by **ds**.

It remains a mystery as to why he did not extend the exercise to other three members of the vector. For when it is done, one gets incomprehensible terms for the three elements individually and for the entire vector as a whole.

The paper, however, does not stop here and maintains that a more complete proof can be presented by considering a system of particles and maintaining conservation of their energy and momentum in stationary as well as moving frames, and before-collision as well as after-inelastic-collision of two equal masses which changed after collision by equal amounts.

The total energy of the system is expressed as sum of the rest energy E_0 and kinetic energy expressed in terms of the first element of the above vector (incorrectly as before). It is then shown that the change in mass due to the inelastic collision is equal to the change in the rest energy of the system. Therefore, the rest energy is the same as mass and vice versa.

The self-prophesy of equating the rest energy with the rest mass in this manner can hardly be accepted. The mistake of arbitrarily applying the unit of c^2 to mass m has also been explained in Observations 2 above.

The Question:

Even if these mistakes are ignored for a moment, how can we apply a mass-kinetic-energy equivalence to all kinds of mass-energy equivalence, starting from nuclear to astronomical levels?

What about the models of Universe based on General Relativity which uses the current rudimentary and mistakes-ridden mass-energy equivalence in its metrics?

We need a derivation that caters to such diverse situations, or declare the relation as only an approximate starting point. This would keep the research and exploration on the right track.

CHAPTER 4

Einstein's Incorrect Derivation of the Lorentz Transformation in his 1916 Book

Introduction:

Einstein, in his 1916 book titled "Relativity: The Special and The General Theory", has presented a derivation of the Lorentz transformation in Annexure I, by kinematics of a light signal moving along the x-axis. The moving frame is taken moving in the (+)ve x-direction.

As will be seen ahead, equations 1 to 5 are written for the moving light signal but the set of equations 5 are applied to rigid bodies like origin of the moving frame and metre-rods placed in the stationary and moving systems. Further, the set of equations 5 are also applied to arbitrary times that do not belong to the moving light signal. Moreover, the equation 3 written for the light signal moving in (-)ve x-direction is incorrect.

The entire derivation is presented below in italics with my intermittent observations placed between two dotted lines in normal font.

Setup & Assumptions:

The derivation is based on the following setup and assumptions:-

i. The distance and time of an event are designated as x and t respectively in stationary frame K and the same are designated as x' and t' in the moving frame K' which is moving with a velocity v with respect to frame K in positive x-direction.

ii. The frames K and K' are aligned in such a way that their coordinate axes all meet at $t = t' = 0$ and that the x and x' axes are permanently aligned.

iii. A light signal is emitted along the positive x-direction at the instant $x = 0$, $t = 0$, $x' = 0$, $t' = 0$, to be observed from both the frames – K as well as K'.

iv. Speed of light in vacuum is the same in all inertial frames.

Observations 1:

Regarding assumptions (iii) & (iv) above, It is important to note that

The object of observation from both the frames is a light signal with its motion equations as $x = ct$ and $x' = ct'$ in positive x-direction. When the relations derived from this assumption are applied to a situation where the object of observation changes to a material body, how can it be acceptable?

The Derivation:

(Note: The main portion of the derivation is quoted)

A light-signal, which is proceeding along the positive axis of x, is transmitted according to the equation

$$x = ct$$

Or,

$$x - ct = 0 \ldots (1)$$

Since the same light-signal has to be transmitted relative to K' with the velocity c, the propagation relative to the system K' will be represented by the analogous formula

$$x' - ct' = 0 \ldots (2)$$

Those space-time points (events) which satisfy (1) must also satisfy (2). Obviously this will be the case when the relation

$$(x' - ct') = \lambda (x - ct). \ldots (3)$$

is fulfilled in general, where λ indicates a constant ; for, according to (3), the disappearance of $(x - ct)$ involves the disappearance of $(x' - ct')$.

Observations 2:

The event (x, t) or (x', t'), as mentioned in 'Setup & Assumptions' above, is for a propagating light signal as confirmed by Eq.(1) & (2). No other type of event can, therefore, conform to these equations and the derivations based on them. Thus the relations based on this derivation cannot be applied to events of other-than-light objects. However, the restriction is flouted by Einstein himself, as shown below and is also being flouted routinely on a widespread scale.

If we apply quite similar considerations to light rays which are being transmitted along the negative x-axis, we obtain the condition

$$(x' + ct') = \mu (x + ct) \ldots (4).$$

Observations 3:

1. The above equation (4) is valid for a case when the event is located on (+)ve x-side and the light signal is moving in (-)ve x-direction, or vice versa, because the two terms x and ct are additive.

   Now, the question is - can the light signal be assumed moving against its own event?; the light signal will never meet its supposed event. So, it is not clear as to what is this statement for. It only shows that the author has something intuitive in his mind which he intends to incorporate in the derivation, even at the cost of correctness.

2. Since the statement of Eq.(4) is also for a light signal similar to Eq.(3), it also attracts the same observations as mentioned in the preceding Observations 2.

By adding (or subtracting) equations (3) and (4), and introducing for convenience the constants a and b in place of the constants λ and μ, where

$$a = \frac{\lambda + \mu}{2}$$

and

$$b = \frac{\lambda - \mu}{2}$$

we obtain the equations

$$\left.\begin{array}{l} x' = ax - bct \\ ct' = act - bx \end{array}\right\} \ldots (5)$$

We should thus have the solution of our problem, if the constants a and b were known. These result from the following discussion. For the origin of K' we have permanently $x' = 0$, and hence according to the first of the equations (5)

$$x = \frac{bc}{a} t$$

If we call v the velocity with which the origin of K' is moving relative to K, we then have

$$v = \frac{bc}{a} \ldots (6)$$

The same value v can be obtained from equations (5), if we calculate the velocity of another point of K' relative to K, or the velocity (directed towards the negative x-axis) of a point of K with respect to K'. In short, we can designate v as the relative velocity of the two systems.

Observations 4:

The set of Equations (5) make statements about motion of a light signal, as observed from the two frames K and K', with certain assumptions already mentioned.

Now, let us examine the statement "For the origin of K' we have permanently x' = 0".

It may be recollected that x' is the distance travelled by the light signal (not the frame K') in a time t'. Then how can the parameter x' be used for the frame K'?

Further, for the light signal, it is never x' = 0 except at the beginning of its travel where we also concurrently have x = 0, t = 0 and t' = 0. All these concurrent values, when substituted in the equations (5), result in zeros on both the sides, which is expected in accordance with the assumptions made.

So, working out of v as above is incorrect.

It is added here that if the likes of equations (5) were not just for light but for any distance-time set, these objections wouldn't have arisen.

Furthermore, the principle of relativity teaches us that, as judged from K, the length of a unit measuring-rod which is at rest with reference to K' must be exactly the same as the length, as judged from K', of a unit measuring-rod which is at rest relative to K.

In order to see how the points of the x'-axis appear as viewed from K, we only require to take a "snapshot" of K' from K; this means that we have to insert a particular value of t (time of K), e.g. t = 0. For this value of t we then obtain from the first of the equations(5)

$$x' = ax$$

Two points of the x'-axis which are separated by the distance $\Delta x' = 1$ when measured in the K' system are thus separated in our instantaneous photograph by the distance

$$\Delta x = \frac{1}{a} \ldots . (7)$$

Observations 5:

As already pointed out in Observations 4, equations (5) are for motion of a light signal with certain assumptions. This fact cannot be ignored while applying any boundary/special condition to the equations. For the light signal at $t = 0$, the other concurrent statements are $x = 0, x' = 0$ and $t' = 0$. By substituting these concurrent values, both sides of the equations become zero, quite in accordance with the assumptions.

Secondly, Eq.(7) has been obtained after application of Δ operator on both sides of the above result i.e. $x' = ax$ and thereby substituting $\Delta x = 1$. Attention is invited to the fact that x' is dependent on x as well as t. We may also say that any relation between $\Delta x'$ and Δx will necessarily involve terms of Δt. Therefore, Δ operator should be applied first, after which a value of Δt can be substituted for any instant t. If Δ operator is applied to the first of equation (5), the result would be as follows

$$\Delta x' = a\Delta x - bc\Delta t$$

For $\Delta t = 0$, this relation would result into $\Delta x' = 0$, as for $\Delta t = 0$, $\Delta x = 0$.

Thus here is another unacceptable deviation from light to meter-rod.

It is reiterated again here that if the likes of equations (5) were not just for light but for any distance-time set, these objections wouldn't have arisen.

But if the snapshot be taken from $K'(t' = 0)$, and if we eliminate t from the equations (5), taking into account the expression (6), we obtain

$$x' = a\left(1 - \frac{v^2}{c^2}\right)x$$

From this we conclude that two points on the x-axis separated by the distance 1 (relative to K) will be represented on our snapshot by the distance

$$\Delta x' = a\left(1 - \frac{v^2}{c^2}\right). \ldots (7a)$$

Observations 6:

Similar to Observations 5.

But from what has been said, the two snapshots must be identical; hence Δx in (7) must be equal to $\Delta x'$ in (7a), so that we obtain

$$a^2 = \frac{1}{1-\frac{v^2}{c^2}} \ldots \ldots (7b)$$

The equations (6) and (7b) determine the constants a and b. By inserting the values of these constants in (5), we obtain the first and the fourth of the equations given in Section 11.

$$\left.\begin{array}{l} x' = \dfrac{x-vt}{\sqrt{1-\frac{v^2}{c^2}}} \\[2ex] t' = \dfrac{t-\frac{v}{c^2}x}{\sqrt{1-\frac{v^2}{c^2}}} \end{array}\right\} \ldots . (8)$$

Thus we have obtained the Lorentz transformation for events on the x-axis. It satisfies the condition

$$x'^2 - c^2 t'^2 = x^2 - c^2 t^2 \ldots . (8a)$$

Observations 7:

Coming to the start of derivation, multiplication of equation (1) with its conjugate and also the same operation for equation (2) will give $x'^2 - c^2 t'^2 = x^2 - c^2 t^2 = 0$. The zero result is obviously for light. For events other than those of light, the result is bound to be non-zero, as for these, $x - ct \neq 0$, $x + ct \neq 0$, $x' - ct' \neq 0$ and $x' + ct' \neq 0$. But these facts have been ignored when switching over to material bodies like the origin of moving frame and unit measuring-rod for finding out value of constants a and b.

The extension of this result, to include events which take place outside the x- axis, is obtained by retaining equations (8) and supplementing them by the relations

$$\left.\begin{array}{l} y' = y \\ z' = z \end{array}\right\} \ldots\ldots (9)$$

In this way we satisfy the postulate of the constancy of the velocity of light in vacuo for rays of light of arbitrary direction, both for the system K and for the system K'. This may be shown in the following manner.

We suppose a light-signal sent out from the origin of K at the time $t = 0$. It will be propagated according to the equation

$$r = \sqrt{x^2 + y^2 + z^2} = ct$$

or, if we square this equation, according to the equation

$$x^2 + y^2 + z^2 - c^2 t^2 = 0 \ldots\ldots (10)$$

It is required by the law of propagation of light, in conjunction with the postulate of relativity, that the transmission of the signal in question should take place — as judged from K1 — in accordance with the corresponding formula

$$r' = ct'$$

Observations 8:

The above method of considering only the **x** value for transformation and leaving untransformed the **y** value and the **z** value, results into a situation where the Lorentz Factor **γ** may take any value, instead of its universally practiced value of $1/\sqrt{1 - \frac{v^2}{c^2}}$. This questions the integrity of the Lorentz transformation itself in cases where the event falls outside the **x**-axis.

It is added here that the same situation arises when the Lorentz transformation, which are derived for events of light ($x/t = c$), are used for events other than those of light, where ($x/t \neq c$).

The same are elaborated below.

For events falling outside the **x**-axis, $r = ct$ and $r' = ct'$, as also mentioned in the derivation.

It entails $x \neq ct$ and $x' \neq ct'$.

Now, the two relations connecting **x** and **x'** are as follows.

$$\left.\begin{array}{l} x' = \gamma(x - vt) \\ x = \gamma(x' + vt') \end{array}\right\} \quad \ldots \ldots (O.1)$$

On multiplying together both the above equations, one gets

$$xx' = \gamma^2 xx' \left(1 - \frac{vt}{x}\right)\left(1 + \frac{vt'}{x'}\right)$$

Or,

$$\gamma = \frac{1}{\sqrt{\left(1 - \frac{vt}{x}\right)\left(1 + \frac{vt'}{x'}\right)}}$$

In case of events of light, since $x = ct$ and $x' = ct'$, the value of γ always works out to

$$\gamma = \frac{1}{\sqrt{1 - \frac{v^2}{c^2}}}$$

which is the familiar Lorentz Factor.

However, in case of events falling outside the **x**-axis, since $x \neq ct$ and $x' \neq ct'$, or for events other than those of light, the value of γ would obviously be different. To work out the value in such cases, therefore, we have to assume it as an unknown and see how it plays out.

With the assumption of **γ** as unknown, let us separate **t** and **t'** one-by-one from the two equations of (O.1). So we get

$$\left. \begin{array}{l} t' = \gamma \left[t - \left(1 - \frac{1}{\gamma^2}\right) \frac{x}{v} \right] \\ t = \gamma \left[t' + \left(1 - \frac{1}{\gamma^2}\right) \frac{x'}{v} \right] \end{array} \right\} \quad \ldots \ldots (O.2)$$

It may be noted that **In case of events of light, an exercise on relations (O.2) in a manner similar to (O.1), fetches us the same standard value of the Lorentz Factor**. However, it is not so in case of other events where $x \neq ct$ and $x' \neq ct'$, as shown below.

Now, to separate γ, we have to multiply together both the relations of either (O.1) or (O.2), and thus we can write as follows.

$$\left. \begin{array}{l} \frac{1}{\gamma^2} = \left(1 - \frac{vt}{x}\right)\left(1 + \frac{vt'}{x'}\right) \\ \frac{1}{\gamma^2} = \left[1 - \left(1 - \frac{1}{\gamma^2}\right)\frac{x}{vt}\right]\left[1 + \left(1 - \frac{1}{\gamma^2}\right)\frac{x'}{vt'}\right] \end{array} \right\} \quad \ldots \ldots (O.3)$$

Note: The first of relations (O.3) should not be used in cases where either $x = 0$ or $x' = 0$. Because in such cases, x or x', as the case may be, disappears from both of the relations (O.1), and the requirement becomes simply $\gamma = t'/t$ or $\gamma = t/t'$ respectively, as can be worked out from relations (O.1) itself. Similarly, the second of relations (O.3) should not be used in cases where either $t = 0$ or $t' = 0$ because in such cases, t or t', as the case may be, disappears from both of the relations (O.2), and the requirement becomes simply $\gamma = x'/x$ or $\gamma = x/x'$ respectively, as can be worked out from relations (O.2) itself.

Now, if we take up any arbitrary value of γ, and find out the values of x' and t', with a given set of x and t, from the first of relations (O.1) and (O.2), and substitute them back into the second of the same relations, we get back the values of x and t we started with. The same is true with interchange of the frames too. After knowing the parameters of both the frames/observers, if we substitute them in relation (O.3), we get back the value of γ we started with.

Thus all values of γ turn out to be conforming to the three relations i.e. (O.1), (O.2) and (O.3).

This is a serious setback to the Lorentz transformation in such cases, as it is not possible to assign unique values to the transformed parameters x' and t', and any set of values become possible for them.

Therefore, the existing scheme of transformation for events falling outside of line of motion of the frame needs to be corrected.

Conclusion:

This is another failed attempt by Einstein, after his 1905 paper, to derive the Lorentz Transformation by kinematics of a moving light signal. Both the works are riddled with mistakes and therefore, are unsuitable for adoption.

To reiterate once again, however, the two failed attempts do not undermine the validity of Lorentz Transformation for light, as these have successfully been derived by electrodynamics and also by kinematics based on reciprocity of velocity of frames, as presented in chapter 1.

The other important problem that needs to addressed is - unsustainability of the proposed scheme of transformation for events lying outside the line of motion of the moving frame/observer. The topic is discussed in detail in chapters 15 and 16.

CHAPTER 5

Spacetime in Question for Permitting Infinite Transformations of an Event

Abstract:

The prerequisite for spacetime is that the length of any of its linear element has to be invariant across all INERTIAL frames in flat (Minkowskian) spacetime and across ALL frames, including accelerated ones, in the curved (Einsteinian) spacetime. The invariance, however, does not lead to unique transformation values for space and time.

Taking the simplest of spacetime, the flat Minkowskian one, for an event (x, t), it means $c^2 t'^2 - x'^2 = c^2 t^2 - x^2 = K$, a constant, for all uniform velocities of an inertial frame (primed) moving in x-direction. The common interpretation is that the condition fetches us different sets of unique values of the transformed distance and time (x', t') at different velocities. However, it is so only when the constant K is zero i.e. for events of light, where one gets the unique values by Lorentz Transformation. When K is more than zero in case of other events, there are infinite combinations of x' and t', for a given event (x, t) and a given frame velocity, that satisfy the condition, as no transformation has been or can be derived for such events. We land up in a state of indeterminacy when trying to work out a transformation for them, like the Lorentz Transformation for light (EM waves), thus permitting infinite number of possible transformations.

The indeterminacy raises serious doubts about the concept itself. It also exposes the flaw in use of the Lorentz Transformations for events that are not of light i.e. where the distance-time ratio is not equal to c. For the same reason, applying them to even the events of light moving in any arbitrary direction, which fall outside the line of motion of the moving frame (say, x-direction), is also incorrect, as the component of the event distance along the line of motion of the frame (x) after a time t is not equal to ct.

Introduction:

Transformations of space and time take place across frames to fulfil the requirement of invariance of length of a linear line element in spacetime. However, there is no tool to find out their individual transformations, unless the spacetime is null i.e. for events of light, where the distance-time ratio is a constant c. The reciprocity of velocity of frames, the essence of Relativity, is the only tool available to work out the transformations separately. However, its use leads to a solution only in case of events of light (null spacetime), and in all other cases (with non-null spacetime), the individual space and time transformations continue to be indeterminate.

To keep the exercise simple, the most common setup of two inertial frames is taken, with their origins coinciding at start and their x-axes aligned parallel to each other. The time at start are also zero in both the frames. The moving frame starts its motion with a uniform velocity v in the positive x-direction. The event (x, t) in the stationary frame falls on the x-axis, which appears as (x', t') in the moving frame. The start event in both the frames is $(0, 0)$. The invariant length of the spacetime linear element from start to the next event is expressed as follows.

$$c^2 t'^2 - x'^2 = c^2 t^2 - x^2 = K$$

where K is a constant with its value as zero for events of light and non-zero for others.

This expression is of the hyperbola that is invariant in both the inertial frames [1]. The events of light fall on the asymptote and other events do on the curved portion of the hyperbola whose vertex is $\sqrt{K}$.

Transformations of Space and Time:

The following fig.1 (I) and (II) may be referred.

The two axes for the spacetime diagram in the frame considered stationary are shown in continuous lines with rectangular orientation and those for the one considered moving are shown in dashed lines with non-rectangular (parallelogram) orientation. [1]

The scales of distance and time in the frame considered stationary are respectively 1 and $1/c$. [1], and those in the frame considered moving have to be worked out, so as to maintain invariance of the spacetime hyperbola for the event.

The primed frame is moving with a velocity v in the (+) ve x-direction with respect to the non-primed frame. So both of its axes are leaning towards the asymptote at an angle δ with respect to their counterparts in the non-primed (stationary) frame [1], as shown in (I).

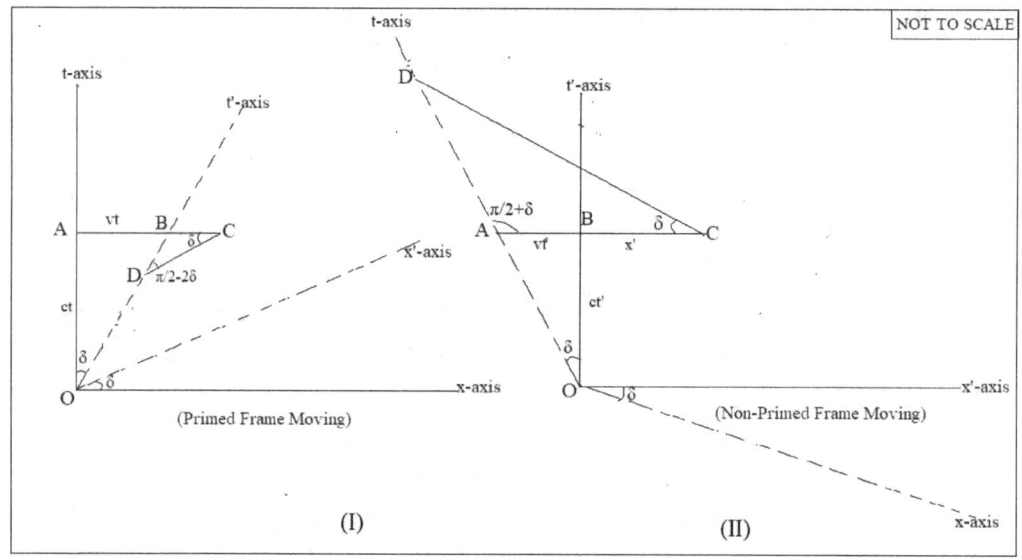

Fig.1

Concurrently, the primed frame may also be considered stationary, with the non-primed one moving, and this is represented by (II). Under this case, the non-primed frame is moving with the same velocity v in opposite direction i.e. (−) ve x-direction. Therefore, both of its axes are leaning away from the asymptote [1] at an angle δ with respect to their counterparts in the primed (stationary) frame, as shown in (II).

The angle δ is dependent on v, and its trigonometric functions are as follows.

$$\left.\begin{array}{l}\tan\delta = \dfrac{vt}{ct} = \beta \\ \cos\delta = \dfrac{1}{\sqrt{1+\beta^2}} \\ \sin\delta = \dfrac{\beta}{\sqrt{1+\beta^2}}\end{array}\right\} \ldots\ldots\ldots\text{Eq. (1)}$$

The point C in both the diagrams represents an event under observation from both the frames.

Thus in Fig.1 (I), the stationary frame ordinates are AC = x, AB = vt and OA = ct, and the moving frame ordinates are DC = x' and OD = ct'.

Similarly in Fig.1 (II), the stationary frame ordinates are BC = x', AB = vt' and OB = ct', and the moving frame ordinates are DC = x and OD = ct.

The axes of the frame considered moving have undergone rotation, as discussed above. This would also necessitate scaling of the axes so that invariance of the event hyperbola is maintained. Though the rotations of the two axes are equal, the required scales along these would be different.

Let the scaling factors for distance and time, in the moving frame coordinates, be denoted as f_x and f_t respectively.

Distance Transformation:

Now, let us take the Fig.1 (I) and apply the Sine Rule in triangle BCD to get

$$\frac{CD}{\sin(\frac{\pi}{2} + \delta)} = \frac{BC}{\sin(\frac{\pi}{2} - 2\delta)}$$

On applying the scale f_x to the moving frame distance ordinate CD, it becomes equal to x'. So, $x' = CD \cdot f_x$. Further, on substituting $BC = x - vt$ and rearranging, we get

$$x' = [x - vt]\frac{\cos\delta}{\cos 2\delta}f_x$$

Or,

$$x' = [x - vt]\frac{\cos\delta}{(\cos\delta)^2 - (\sin\delta)^2}f_x$$

On using the relations of Eq. (1) we get

$$x' = [x - vt]\left[\dfrac{1/\sqrt{1+\beta^2}}{(1-\beta^2)/(1+\beta^2)}\right]f_x$$

Or,

$$x' = [x - vt]\dfrac{\sqrt{1+\beta^2}}{(1-\beta^2)}f_x$$

Or,

$$x' = x\left[1 - \dfrac{vt}{x}\right]\dfrac{\sqrt{1+\beta^2}}{(1-\beta^2)}f_x \quad \ldots\ldots\ldots \text{Eq. (2)}$$

Let us now take the Fig.1 (II), apply the Sine Rule in triangle ACD and also the scale f_x to the moving frame distance ordinate DC, as explained above. On proceeding further in a manner similar to the preceding exercise, we get

$$x = x'\left[1 + \dfrac{vt\prime}{x\prime}\right]\dfrac{\sqrt{1+\beta^2}}{(1-\beta^2)}f_x \quad \ldots\ldots\ldots \text{Eq. (3)}$$

The distance transformation relation now boils down to finding the value of f_x. On multiplying eq. (2) and eq. (3), and separating f_x, we get

$$f_x = \dfrac{1}{\sqrt{\left(1 - \dfrac{vt}{x}\right)\left(1 + \dfrac{vt'}{x'}\right)}}\dfrac{(1-\beta^2)}{\sqrt{1+\beta^2}}$$

This will be discussed after the expression for time transformation is also worked out below.

Time Transformation:

In fig.1 (I), the value of BD in the moving frame may be obtained by applying the Sine Rule in triangle BCD, which is as follows

$$\frac{BD}{\sin \delta} = \frac{BC}{\sin(\frac{\pi}{2} - 2\delta)}$$

Or,

$$BD = \frac{\sin \delta}{1 - 2(\sin \delta)^2}(AC - BC)$$

Further, we have

$$OD = OB - DB$$

So,

$$OD = \left[OB - \frac{\sin \delta}{1 - 2(\sin \delta)^2}(AC - BC)\right]$$

Now, on substituting the values of OD, OB, AC and BC in above equation and also applying the Scale f_t to the moving frame time ordinate OD, as explained before, we get

$$ct' = \left[\frac{vt}{\sin \delta} - \frac{(x - vt)\sin \delta}{1 - 2(\sin \delta)^2}\right]f_t$$

On using the relations of Eq. (1), we get

$$ct' = \left[ct\sqrt{1 + \beta^2} - \frac{(x - vt)\beta\sqrt{1 + \beta^2}}{[1 + \beta^2 - 2\beta^2]}\right]f_t$$

Or,

$$ct' = \left[ct - \frac{(x - vt)\beta}{[1 + \beta^2 - 2\beta^2]}\right]\sqrt{1 + \beta^2}f_t$$

Or,

$$ct' = \left[ct(1 - \beta^2) - (x - vt)\frac{v}{c}\right]\frac{\sqrt{1 + \beta^2}}{(1 - \beta^2)}f_t$$

Or,

$$ct' = \left[ct - \frac{v}{c}x\right] \frac{\sqrt{1+\beta^2}}{(1-\beta^2)} f_t$$

Or,

$$t' = t\left[1 - \frac{v}{c^2}\frac{x}{t}\right] \frac{\sqrt{1+\beta^2}}{(1-\beta^2)} f_t \quad \ldots\ldots\ldots \text{Eq. (4)}$$

Also from fig.1 (II), the value of AD in the moving frame may be obtained by applying the Sine Rule in triangle ACD and the scale f_t to the moving frame time ordinate OD.

On proceeding further in a manner similar to the just completed exercise above, we get the following.

$$t = t'\left[1 + \frac{v}{c^2}\frac{x'}{t'}\right] \frac{\sqrt{1+\beta^2}}{(1-\beta^2)} f_t \quad \ldots\ldots\ldots \text{Eq. (5)}$$

On multiplying eq. (4) and eq. (5), and separating f_t, we get

$$f_t = \frac{1}{\sqrt{\left(1 - \frac{v}{c^2}\frac{x}{t}\right)\left(1 + \frac{v}{c^2}\frac{x'}{t'}\right)}} \frac{(1-\beta^2)}{\sqrt{1+\beta^2}}$$

Similar to the distance transformation, the relation for the time transformation also gets stuck at finding the solution of f_t.

Indeterminacy of f_x and f_t raises serious questions about spacetime:

The two expressions of f_x and f_t obtained above are put together as follows.

$$\left.\begin{array}{l} f_x = \dfrac{1}{\sqrt{\left(1-\frac{vt}{x}\right)\left(1+\frac{vt'}{x'}\right)}} \dfrac{(1-\beta^2)}{\sqrt{1+\beta^2}} \\[2em] f_t = \dfrac{1}{\sqrt{\left(1-\frac{v\,x}{c^2 t}\right)\left(1+\frac{v\,x'}{c^2 t'}\right)}} \dfrac{(1-\beta^2)}{\sqrt{1+\beta^2}} \end{array}\right\} \ldots\ldots (6)$$

An examination of the expressions of f_x and f_t reveals that for a given event (x, t) and frame velocity v, infinite number of values f_x and f_t are possible, depending on what values of the transformed distance and time (x', t') one selects, subject, of course, to the relation $c^2 t'^2 - x'^2 = c^2 t^2 - x^2$. This is also true the reverse way.

That is because for a given set of x, t and v, the values of f_x and f_t depend on x' and t' which in turn are dependent on f_x and f_t respectively. Similarly, for a given set of x', t' and v, the value of f_x and f_t depend on x and t which in turn are dependent on f_x and f_t respectively.

On the other hand, if one starts with some assumed values of f_x and f_t, one may calculate the transformations of distance and time with eq. (2) to (5). However, the question remains as to which value of theirs is the correct one, and there is no answer.

Thus f_x and f_t continue to be indeterminate.

Events of Light do not Create Indeterminacy:

There is an exception, however. The events of light have null spacetime and do not create such indeterminacy. Such events fall on the asymptote of the spacetime hyperbola. In the instant case, the event point C falls on the asymptote which makes an angle of $\pi/4$ with both the rectangular axes, in each of the two the frames when the concerned frame is considered stationary.

For events of light, $x/t = x'/t' = c$. On substitution of the ratios x/t and x'/t' with c, both the expressions of eq. (6) for the scales reduce to a single expression as follows.

$$\left. \begin{array}{l} f_x = \sqrt{\dfrac{1-\beta^2}{1+\beta^2}} \\[6pt] f_t = \sqrt{\dfrac{1-\beta^2}{1+\beta^2}} \\[6pt] f_x = f_t \end{array} \right\} \quad \ldots \ldots (7)$$

Now, the common factor appearing on extreme right in eq. (2) to (3) is $\dfrac{\sqrt{1+\beta^2}}{(1-\beta^2)} f_x$ and that in in eq. (3) to (4) is $\dfrac{\sqrt{1+\beta^2}}{(1-\beta^2)} f_t$.

On substitution of the above expressions of f_x and f_t, both the common factors become $1/\sqrt{1-\beta^2}$ which is the Lorentz Factor γ, and the eq. (2) to (5) become the familiar Lorentz Transformation for events of light.

Coming back to the events (distance-time sets) other than those of light, there are no unique values of f_x and f_t, and infinite number of transformations are possible. Thus the spacetime hyperbola of the stationary frame may appear in infinite number of ways in a single moving frame, and this shatters the concept of invariant spacetime hyperbola, or invariant lengths of spacetime linear elements, across frames.

How does the Indeterminacy go Unnoticed?

We have just now seen that it is only for events of light that the eq. (6) for the unknown scaling factors f_x and f_t get a common solution that results into emergence of the Lorentz Factor γ in the transformation relations (2) to (5).

However, the same solution/transformations, with the Lorentz Factor γ, are used even for events that are not of light, i.e. even for arbitrary distance-time sets. Use of the Lorentz Factor γ means that the values of f_x and f_t are according to the eq. (7). These values do not conform to eq. (6), which are derived for such events to strike conformity with the prerequisite of spacetime.

Thus what goes unnoticed in such cases is non-conformity of the utilized values of f_x and f_t with eq. (6), thus violating the very prerequisite for spacetime.

Corroboration from another Approach:

The spacetime equation/condition is

$$c^2 t'^2 - x'^2 = c^2 t^2 - x^2 = K$$

Where K is a constant. Now, let n be a real number.

If $ct' = nct$, then x' should be as follows to satisfy the above condition

$$x' = \sqrt{n^2 x^2 + (n^2 - 1)K}$$

Similarly, if $x' = nx$, then ct' should be as follows to satisfy the above condition

$$ct' = \sqrt{n^2 c^2 t^2 - (n^2 - 1)K}$$

The number n may take infinite values while satisfying the above condition. Thus there are infinite number of combinations of x' and ct' that will satisfy the condition, with just one value of K and the velocity of the moving frame v being nowhere in picture. The infinite number of combinations of (x', t') for a given set of (x, ct) and v correspond to the infinite number of values of f_x and f_t.

Events of Light Falling Outside the Line of Motion of the Moving Frame:

So far, we have shown that the eq. (6) get a solution for events of light, resulting into the Lorentz Transformation. However, this is true only for those falling on the line of motion of the moving frame, and not for those that fall outside the aforesaid line, even for events of light. The reason is explained below.

In the given setup, as described in para 2 of Introduction, if the light signal was moving in any arbitrary direction other than the x-direction, it would generate events that fall outside the line of motion of the moving frame. For such events, the x-component of their distance from the origin is to be taken, which is not equal to c times the time elapsed. That effectively makes them events that are not of light, as already explained.

Conclusion:

The above simple exercise on flat spacetime shows how readily the Lorentz Transformation derive itself for events of light, while indeterminacy is encountered in case of other events, making possible infinite number of transformations between two inertial frames, for a given event. Thus it challenges the cardinal feature of spacetime, i.e. invariance of its line element across all frames, even in the simplest case of flat spacetime with inertial frames, leave aside the curved spacetime with accelerated frames.

It also establishes that it is incorrect to use the Lorentz Transformations for events that are not of light i.e. where the distance-time ratio is not equal to c. For the same reason, applying them to even the events of light moving in any arbitrary direction, which fall outside the line of motion of the moving frame (say, x-direction), is also incorrect, as the component of the event distance along the line of motion of the frame (x) after a time t is not equal to ct.

References:

Space and Time (Minkowski's Papers on Relativity), Translated by Fritz Lewerto and Vesselin Petkov, Edited by Vesselin Petkov, Free Version, from Minkowski Institute Press.

INCORRECT APPLICATIONS

(OF INVALID LORENTZ TRANSFORMATION)

CHAPTER 6

The Unsound Structure of Relativity

Abstract:

The Lorentz Transformation Condition, and its spinoff, the Lorentz Factor are the founding blocks of Relativity. At the same time, the reciprocity is inherent to, and inseparable from Relativity, which states that either of the two frames in relative motion could be considered stationary, with the other one as moving. Therefore, it becomes mandatory for the theory of Relativity to simultaneously strike conformity with all the three. A closer look, however, reveals that although the requirement is met for events of light, the initial agent considered for promulgation of the theory, the discordance is far and wide in respect of other/arbitrary events (distance-time sets), which form a vast majority. Thus it raises a serious question about the validity of the theory, in its present form, when it comes to its application in majority of the cases. The spacetime, and other theories based on it, also come into serious question. The article discusses such areas of inconsistency.

Introduction:

The proverbial chinks in the armour of a theory can be gauged by the self-contradictions it harbours, and also by the disparity in results it shows up, when approached in different ways.

The following brief history of its evolution throws light on how the unsoundness has crept in.

From Maxwell's wave equation for EM waves/**light**, Lorentz inferred an expression $(c^2 t^2 - x^2)$, with its value obviously as zero, which remained invariant in all inertial frames moving **co-directionally with the (distance traversed by) light**. The theme was picked up by Einstein in 1905 to derive Lorentz transformation by kinematics. However, the transformation so derived had to remain limited to the space and time traversed only by light.

Minkowski in 1908 took a non-zero value of the expression, and that too had to remain invariant **under Lorentz boosts, co-directional with the distance from origin**, as the invariance was for the expression, without any reference to its numerical value.

No questions were raised, as the change of value from zero to non-zero did not affect conformity to the Lorentz transformation, under the same conditions.

What, however, got surrendered was the reciprocity which is inherent to, and inseparable from the very concept of Relativity, which was expected to expand its reach, with the above modification, to cover all events (distance-time sets).

This indifference, or ignorance, inducted contradictions/unsoundness in the theory, which even now, we are thriving on and building new theories upon.

Instances of Self-Contradictions:

Such instances are discussed below item-wise.

1. Lorentz Factor not Compatible with Reciprocity, for Arbitrary Events (Other Than Those of Light)!

The Lorentz transformation relations for the simple case practiced today i.e. when the direction of motion of the observer/frame is parallel to that of the event's location from the origin, are as follows [1].

$$\left. \begin{array}{l} x' = \gamma(x - vt) \\ t' = \gamma(t - vx/c^2) \end{array} \right\} \ \ldots \ (1)$$

$$\left. \begin{array}{l} x = \gamma(x' + vt') \\ t = \gamma(t' + vx'/c^2) \end{array} \right\} \ \ldots \ (2)$$

Where $\gamma = 1/\sqrt{1 - v^2/c^2}$

It does not need over emphasis that the relations (1) and (2) together display the essence of Relativity i.e. reciprocity, which means that one can interchange the motion states of the primed and the non-primed frames, by replacing v with $-v$, without changing the structure of the transformation relations.

The expression for γ may be noted, and at the same time, it may also be recalled that the same relations are used, without any riders, for all events or distance-time pairs, not just limited to those of light.

Now, let us bring together the relations of distance and time separately as follows.

$$\left.\begin{aligned} x' &= \gamma x(1 - vt/x) \\ x &= \gamma x'(1 + vt'/x') \end{aligned}\right\} \quad \ldots\ldots (3)$$

$$\left.\begin{aligned} t' &= \gamma t(1 - vx/c^2 t) \\ t &= \gamma t'(1 + vx'/c^2 t') \end{aligned}\right\} \quad \ldots\ldots (4)$$

On multiplying together the two relations of (1) and (2) separately, one-by-one, and separating γ, we get the following two expressions which are equivalent.

$$\left.\begin{aligned} \gamma &= \frac{1}{\sqrt{\left(1 - \frac{vt}{x}\right)\left(1 + \frac{vt'}{x'}\right)}} \\ \gamma &= \frac{1}{\sqrt{\left(1 - \frac{vx}{c^2 t}\right)\left(1 + \frac{vx'}{c^2 t'}\right)}} \end{aligned}\right\}$$

For events of light, since $x = ct$ and $x' = ct'$, both the above expressions reduce to the one practiced today i.e. $\gamma = 1/\sqrt{1 - v^2/c^2}$.

However, in case of other events, forming the majority, since $x \neq ct$ and $x' \neq ct'$, none of the expressions reduce to what is practiced today, and the larger the divergence between the distance-time ratio of the event from c, the larger is the mistake in adoption of the current Lorentz Factor.

The result, therefore, also questions the validity of application of Relativity, in its current form, to Particle Physics, Quantum Physics, General Relativity and Spacetime etc.

2. Observers Moving Oblique to Direction of Light Do Not Find the Transformation Conforming to Reciprocity, with the Current Lorentz Factor!

When the observer is moving along x-axis and the light signal in any arbitrary direction in 3D space, the current practice [1] is to transform only the component of distance travelled by the signal in x-direction, along with the entire time in accordance with the relations (1) and (2) above, and leave the other components untransformed.

It is practiced universally without paying attention to the fact that it contradicts the reciprocity itself, which is a mandatory requirement of Relativity. Let us see how.

It may be recalled that in this case, $ct = \sqrt{x^2 + y^2 + z^2}$ and $ct' = \sqrt{x'^2 + y^2 + z^2}$. Resultantly, one may also state that $x \neq ct$ and $x' \neq ct'$.

With these relations, as already shown in the previous sub-section, none of the two expressions of γ, as arrived at, reduces to $\gamma = 1/\sqrt{1 - v^2/c^2}$, which otherwise always holds in case of events of light, and which is used in transformation of x.

It thus raises an alarm about mistakes in the current practice.

The **solution lies in** deriving a new transformation [2] of distance and time, based on the direct distance from the moving frame to the light signal which, although being oblique to the direction of motion of the frame, as well as to that of the light signal in the stationary frame, conforms to the postulate i.e. invariance of light speed in all inertial frames.

3. Events of Light 'Do' and 'Do Not' Conform to the Lorentz Transformation Condition!

Such a peculiar result arises because for such events, the distance and time maintain the same ratio in all inertial frames, and the transformation relations are also structured on the same requirement.

To see this phenomenon, we recall the starting part of the simplified derivation presented by Einstein in his 1916 book [1], which is placed below in italics, with dotted lines at beginning and the end.

A light-signal, which is proceeding along the positive axis of x, is transmitted according to the equation

$$x = ct$$

Or,

$$x - ct = 0 \ldots (1)$$

Since the same light-signal has to be transmitted relative to K' with the velocity c, the propagation relative to the system K' will be represented by the analogous formula

$$x' - ct' = 0 \ldots (2)$$

Those space-time points (events) which satisfy (1) must also satisfy (2). Obviously this will be the case when the relation

$$(x' - ct') = \lambda (x - ct) \ldots (3)$$

is fulfilled in general, where λ indicates a constant ; for, according to (3), the disappearance of $(x - ct)$ involves the disappearance of $(x' - ct')$.

If we apply quite similar considerations to light rays which are being transmitted along the negative x-axis, we obtain the condition

$$(x' + ct') = \mu (x + ct) \ldots (4).$$

By adding (or subtracting) equations (3) and (4), and introducing for convenience the constants a and b in place of the constants λ and μ, where

$$a = \frac{\lambda + \mu}{2}$$

and

$$b = \frac{\lambda - \mu}{2}$$

we obtain the equations

$$\left.\begin{array}{l} x' = ax - bct \\ ct' = act - bx \end{array}\right\} \ldots (5)$$

Moving ahead through the derivation, the expressions of a and b have been worked out as follows.

$$a = 1/\sqrt{1 - v^2/c^2}$$

$$b = (v/c)/\sqrt{1 - v^2/c^2}$$

On substitution of the above expressions in relation (5) of the above quote, we get the Lorentz transformation as shown below.

$$\left.\begin{array}{l} x' = \gamma(x - vt) \\ t' = \gamma(t - vx/c^2) \end{array}\right\}$$
$$\left.\begin{array}{l} x = \gamma(x' + vt') \\ t = \gamma(t' + vx'/c^2) \end{array}\right\}$$

Where $\gamma = a = 1/\sqrt{1 - v^2/c^2}$

Conformity to Lorentz Transformation Condition-

In the text quoted above, it is obvious that $\lambda = (a + b)$ and $\mu = (a - b)$. With these values, the relations (3) and (4) of the quote, which are stated in the form of the combined terms of distance and time, may be expressed as follows.

$$(x' - ct') = \sqrt{\frac{1+v/c}{1-v/c}}(x - ct)$$

$$(x' + ct') = \sqrt{\frac{1-v/c}{1+v/c}}(x + ct)$$

On multiplying together the above two relations, we get

$$c^2 t'^2 - x'^2 = c^2 t^2 - x^2$$

Which is in conformity to the Lorentz Transformation Condition.

Further, even when one substitutes the Lorentz transformation expressions of x' and t' on the LHS of the condition i.e. $(c^2 t'^2 - x'^2)$, one gets the RHS of the condition i.e. $(c^2 t^2 - x^2)$, provided $\gamma = 1/\sqrt{1 - v^2/c^2}$, which is true only for events of light.

Non-Conformity to Lorentz Transformation Condition-

Now, on substituting $x = ct$ and $x' = ct'$ in the above four Lorentz transformation relations, and solving further with the terms of γ, we get the following individual relations for distance and time in the two frames.

$$x' = \sqrt{\frac{1-v/c}{1+v/c}}\, x \quad \text{and} \quad t' = \sqrt{\frac{1-v/c}{1+v/c}}\, t$$

$$x = \sqrt{\frac{1+v/c}{1-v/c}}\, x' \quad \text{and} \quad t = \sqrt{\frac{1+v/c}{1-v/c}}\, t'$$

When the above expressions are substituted on LHS of the Lorentz Transformation Condition, we get

$$c^2 t'^2 - x'^2 = \left(\frac{1-v/c}{1+v/c}\right)(c^2 t^2 - x^2)$$

Which is **not** in conformity with the Lorentz Transformation Condition, although the values on both the sides are zero.

It is noteworthy that the **transformation factor for the combined term $(x + ct)$ is the same as that for x and t individually. However, the transformation factor for the combined term $(x - ct)$ is inverse of that for the combined term $(x + ct)$, as well as that for x and t individually** .

Thus one gets two different results from two approaches.

Conclusion:

The failure of the current theory to sustain its own essence i.e. reciprocity, in respect of events other than those of light, is very serious. It calls for a total recast of the theory, rather than building on a base created for events of light.

Going by the exercise carried out in item no.1 above, it is almost impossible to arrive at a unique Lorentz Factor which would support reciprocity in case of all arbitrary events. It also means the current theory is good only for the events of light. Limiting the theory to only light would make the dependent theories like spacetime and General Relativity etc. redundant.

Need for a new theory is already felt, for objects other than light, in view of the time dilation and length contraction reportedly occurring in moving bodies, in contradiction to the very fact that even moving frames are stationary to themselves, and therefore, motion was of no consequence within their own frame.

As regards the item no.2 discussed above, working out a new transformation [2] on the entire oblique distance from the moving frame to light signal, rather than applying the existing transformation on a component of distance, removes the contradiction.

References:

1. A. Einstein's Book "Relativity: The Special and the General Theory", 1916.
2. Authouri's Book "Refining Relativity Part 1 (The Special Theory)", 2020.

CHAPTER 7

Manoeuvring of Results in Relativity by Manipulating Inputs

Introduction:

The Lorentz Transformation have been derived only for events of light and cannot be derived for any other arbitrary event i.e. distance-time set. However, these are used extensively for all kinds of events i.e. distance-time sets to explain away many physical phenomena.

While the events of light offer a limited freedom to choose inputs of distance and time for transformation, bound by $x = ct$, others offer enormous flexibility for manipulation, depending on what we want to show/experience. In other words, when the Lorentz Transformation valid only for distance-time sets bound by $x = ct$ are force-applied to those not honouring the said condition, any manipulations are possible to show the desired result.

As already pointed out, even Einstein did it extensively. The article shows how bizarre the results of transformation could be by selective inputs of distance and time.

Source & Constraints:

The above statement emanates from the Lorentz transformation relations which are as follows.

$$x' = \frac{x - vt}{\sqrt{1 - \frac{v^2}{c^2}}}$$

$$t' = \frac{t - \frac{v}{c^2}x}{\sqrt{1 - \frac{v^2}{c^2}}}$$

As usual, the primed parameters are for the moving frame and the non-primed for the stationary one. All parameters carry their usual meaning. The velocity of observer v is in the direction of increasing x.

It is important to remember the constraint that the above relations are true only for a case where the frame/observer is moving parallel to the distance x, and therefore, the entire exercise ahead is in the same context only.

Further, selection of arbitrary distance-time pairs at one's choice is not possible for light events, as the pair has necessarily to adhere to the ratio c. For such events, the only manipulation that is possible is the change of direction of motion of the observer.

The Exercise:

Let there be two events 1 and 2 in sequence, distance-wise as well as time-wise, and parallel to the direction of motion of the observer. Let us first write the above relations separately for the two and take their differences as follows, showing the distance and time intervals in the two frames.

$$x_2' - x_1' = \frac{(x_2 - x_1) - v(t_2 - t_1)}{\sqrt{1 - \frac{v^2}{c^2}}}$$

$$(t'_2 - t'_1) = \frac{(t_2 - t_1) - \frac{v}{c^2}(x_2 - x_1)}{\sqrt{1 - \frac{v^2}{c^2}}} \quad \ldots\ldots \text{Eq.(1)}$$

In both the expressions, the common denominator on the RHS is $\sqrt{1 - \frac{v^2}{c^2}}$ which is always less than 1 and therefore, would always have the effect of increasing the value of distance or time interval, when measured by the moving observer.

However, the second terms in both the numerators are subtractive, and therefore, would have the effect of decreasing the values.

Obviously, the net effect would determine as to what the moving observer infers. Thus both – contraction as well as elongation/expansion – of distance and time could occur and therefore, no general inference like length contraction or time dilation could be drawn as yet by the moving observer.

Now, let us examine as to how to draw general/universal conclusions.

EVENTS OF LIGHT

For events of light, the relations of eq.(1) reduce to the following on substituting $t = x/c$ in the first and $x = ct$ in the second.

$$\left.\begin{array}{l} x_2' - x_1' = \sqrt{\dfrac{1 - v/c}{1 + v/c}}(x_2 - x_1) \\[2ex] t_2' - t_1' = \sqrt{\dfrac{1 - v/c}{1 + v/c}}(t_2 - t_1) \end{array}\right\}$$

Since there is no option for choosing arbitrary distance-time sets, there is always a contraction, of length and time traversed by light, for observers moving in the direction of light.

However, what happens to observers moving in a direction opposite to that of the observed light? Obviously, with v now becoming negative, there would always be elongation.

So, the general statement for events of light is as follows.

There is always a contraction of length and time traversed by light for observers moving in its direction, and the observers moving in opposite direction would always find the reverse i.e. elongation.

OTHER-THAN-LIGHT EVENTS

Since there is no restriction on such distance-time sets (events) to conform to some ratio, manipulations are possible to a large extent. The same are discussed below.

To reiterate, all the effects/results are as referred from the moving frame, and the manipulations are done only from the stationary frame where the objects of transformation lie.

The manipulations may be classified in two groups i.e. sign-lead and value-lead. The former is of a general nature and targets only the sign of transformation (contraction or elongation/expansion), whereas the latter is capable of fetching the result with the desired sign and magnitude.

A. Sign-Lead Manipulations:

These are the manipulations of a general nature in stationary frame, involving no calculations, and are carried out by either culling or reversal of sign of the intervals of distance or time, to get the desired result.

As we will see ahead, some of them are easily available, and others have specifically to be created.

A.1. Elongation-Lead Manipulations:

These manipulations are directly done with the terms on the RHS of eq.(1), which belong to the stationary frame carrying the objects of transformation. The numerators, carrying three parameters, offer a wide scope of general manipulations that surely lead to elongation, by making the two terms additive, as the denominator always causes elongation. These are discussed below.

[**Note:** For the same reason of denominator's role, it not possible to design a general type of manipulation from the stationary frame, which would surely lead to contraction. It requires either imposition of a general condition for seeing results in the moving frame, or calculations, as we will see in sections ahead.]

1. **Sure Elongation of Length:**

    Refer the first of eq.(1).

1.1. By Freezing Time at the Ends -

If the second term of the numerator is made zero by freezing (selecting the same) time at both the ends in the stationary frame, there would always be length elongation in the moving frame.

Such a condition can be simulated by a setup where the ends of a length light up simultaneously in the stationary state, and the moving observer measures the locations of the lights (appearing at different times for him/her) and works out the length by taking difference of the two measurements.

Einstein, in his book "Relativity: The Special and The General Theory", Chapter 12 (The Behaviour of Measuring-Rods and Clocks in Motion), has undertaken this step to show that a measuring rod of length $\sqrt{1 - v^2/c^2}$ in the stationary frame as seen in a snapshot (both end times taken as zero) measures 1 in the moving frame, but while stating the result, he erroneously says that a moving rod of length 1 contracts to $\sqrt{1 - v^2/c^2}$. The details are given in chapter 11.

He should have utilized the conditions mentioned in section A.2.(3) ahead to establish length contraction in motion always.

The correct statement should be – "By freezing time at ends of a rod in stationary frame, one would always see it elongated in the moving frame."

1.2. By Doing *EITHER* of the Following -

1.2.1 By Reordering Times at the Ends –

The second term of the numerator can be made additive by making $(t_2 - t_1)$ as $(-)$ve i.e. by selecting $(t_2 < t_1)$.

Taking the previous example, it can be made to happen by lighting up end 2 first, and end 1 subsequently.

1.2.2 By Reversing the Direction of Observer's Motion –

The second term of the numerator can also be made additive by making v as $(-)$ve i.e. by reversing the observer's direction of motion.

1.2.3 By Reordering Locations of the Ends -

The two terms of the numerator can also be made additive by making $(x_2 - x_1)$ as (-)ve i.e. by ensuring $(x_2 < x_1)$.

This can be made possible by selecting the event 2 in a direction opposite to that of motion of the observer, or in (-)ve direction, with respect to event 1.

2. Sure Expansion of Time Interval (Time Dilation):

Refer the second of eq.(1).

2.1. By Freezing Location at the Ends of Time Interval (Start and Finish) -

Similar to the case 1 above, if the second term of the numerator is made zero by selecting the same distance at both the ends of time (start and finish) in the stationary frame, there would always be time elongation/expansion in the moving frame.

This has been **undertaken by Einstein,** in the same chapter of his book as mentioned in 1.1 above, to show time dilation by taking the same location (origin of frame K' which is stationary for the clock) for the two ticks of the clock.

2.2. By Doing *EITHER* of the Following –

2.2.1. By Reordering Locations of the Ends of Time Interval –

The two terms of the numerator can also be made additive by making $(x_2 - x_1)$ as (-)ve i.e. by ensuring $(x_2 < x_1)$.

This can be made possible by selecting the event 2 in a direction opposite to that of motion of the observer, or in (-)ve direction, with respect to event 1.

2.2.2. By Reversing the Direction of Observer's Motion –

The second term of the numerator can also be made additive by making v as (-)ve i.e. by reversing the observer's direction of motion.

2.2.3. By Reordering Times at the Ends –

The two terms of the numerator can be made additive by making $(t_2 - t_1)$ as (-)ve i.e. by ensuring $(t_2 < t_1)$.

Taking the example of 1.1, it can be made to happen by lighting up end 2 first, and end 1 subsequently.

A.2. Contraction-Lead Manipulations:

As mentioned earlier, these general manipulations are also done without any calculation and are in the form of a condition in the moving frame, while observing results, by freezing either time at the two ends of a length, or location at the two ticks (start and finish) of a time interval i.e. by manipulating LHS of eq.(1). The same are discussed below.

3. Sure Contraction of Length by Freezing Times at the Ends:

It happens when the observer measures the length in a snapshot (of his frame) i.e. $(t'_2 - t'_1) = 0$.

It means LHS of the second relation of eq.(1) becomes zero, which leads to

$$(t_2 - t_1) = \frac{v}{c^2}(x_2 - x_1)$$

This is a demand on the interval of time to be selected between the two ends in the stationary frame.

On substituting the expression of $(t_2 - t_1)$ in the first of eq.(1), one gets

$$(x'_2 - x'_1) = (x_2 - x_1)\sqrt{1 - v^2/c^2}$$

This means a sure contraction of length always.

However, such a **statement comes with a hidden restriction** on the time interval to be selected in stationary frame, which is not mentioned by anybody.

4. Sure Contraction of Time by Freezing Locations at its Ends (Start and finish):

It would happen if the observer chose to record the start and finish of time from the same location in his frame. This was obviously possible with two synchronised clocks placed at two different locations with the specified distance interval in the stationary frame, as worked out below.

The condition means LHS of the first relation of eq.(1) becoming zero, which leads to

$$(x_2 - x_1) = v(t_2 - t_1)$$

This gives the separation required for the two synchronizes clocks.

On substituting the expression of $(x_2 - x_1)$ in the second relation of eq.(1), we get

$$(t'_2 - t'_1) = (t_2 - t_1)\sqrt{1 - v^2/c^2}$$

Which means a sure contraction of time always.

B. Value-Lead Manipulations:

These manipulations are intended to achieve a transformation of the desired sign and magnitude, and thus include all the Sign-Led Manipulations discussed in the preceding section. Obviously, therefore, calculations become necessary. Since the numerators of eq.(1) contain subtractive terms causing decrease, and the denominator always causes increase, there is a seesaw battle between the two for determining the final result.

Therefore, we have to examine their relative roles in the expressions, and to achieve our objective, what could be a better starting point than working out of the pivot of the seesaw. Technically, one may define the pivot as a set of observer's velocity and one of the event variables (distance or time interval) which results into no-transformation of its co-variable (time or distance interval respectively).

No-transformation of length and time interval respectively mean

$$(x_2' - x_1' = x_2 - x_1) \text{ and } (t_2' - t_1' = t_2 - t_1)$$

On substituting these in the first and the second relations respectively of eq.(1) and rearranging, one gets the corresponding time interval, say $(t_2 - t_1)_{nt}$ and distance interval, say $(x_2 - x_1)_{nt}$ which would cause no transformation of $(x_2 - x_1)$ and $(t_2 - t_1)$ respectively. These are shown below.

$$(t_2 - t_1)_{nt} = \frac{(x_2 - x_1)}{c}\left[\frac{c}{v}\left(1 - \sqrt{1 - \frac{v^2}{c^2}}\right)\right]$$

$$(x_2 - x_1)_{nt} = c(t_2 - t_1)\left[\frac{c}{v}\left(1 - \sqrt{1 - \frac{v^2}{c^2}}\right)\right]$$

There is a common factor in brackets in both the above relations. If we denote this term with A, the above relations reduce to

$$(t_2 - t_1)_{nt} = \frac{(x_2 - x_1)}{c} A$$
$$(x_2 - x_1)_{nt} = c(t_2 - t_1) A$$

The value of A is dependent on the speed of the observer v and that of light in vacuum c, and is always less than 1. For example, $A = 0.724$ for $v = 0.95c$ and $A = 0.025$ for $v = 0.05c$.

It may be noted that the above values of time and distance intervals, though causing no-transformation of their co-variables, themselves transform so much that their value becomes of opposite sign, with the same magnitude. These can be worked out by substituting the above expressions in the first and the second relations respectively of eq.(1). The results are shown below.

$$(t'_2 - t'_1)_{nt} = -(t_2 - t_1)_{nt}$$
$$(x'_2 - x'_1)_{nt} = -(x_2 - x_1)_{nt}$$

Coming to the manipulations, it now needs no explanation that

1. There would be contraction or elongation of a length, depending on whether the interval between the times at its ends was respectively more or less than the No-transformation time interval $(t_2 - t_1)_{nt}$.

2. There would be contraction or elongation of a time interval, depending on whether the interval between the locations of its ends (start and finish) was respectively more or less than the No-transformation distance interval $(x_2 - x_1)_{nt}$.

It is also clear that the Sign-Led Manipulations, which involve setting the distance interval or the time interval to zero or less than zero, are a subset of the Value-Led Manipulations.

Conclusion:

The above exercise is an eye-opener for many, especially those who have got seasoned to the theme of universal 'length contraction' and 'time dilation' in moving frames. It is a reminder that the hidden conditions have to be borne in mind.

It is emphasised that all the transformations have emanated from the existing Lorentz transformation relations, as being done today, despite NOT being valid for these arbitrary other-than-light events.

For events of light, only one parameter i.e. observer's velocity v, is available to carry out manipulations, to change the sign and/or value of transformation. The transformations of light-traversed segments of distance and time change from contraction to elongation, when the observer reverses his/her motion from being co-directional to be anti-directional, with respect to the light ray/signal. This, though having a limited use in worldly situations, is very important in astronomical observations.

Other distance-time sets, not bound by any ratio constraint, come with enormous freedom for manipulations, to get a transformation of the desired sign and value, at a particular velocity of the observer. The no-transformation values provide a fulcrum on which the transformation signs (contraction vs. elongation) seesaw.

So, remember that the transformation effects on physical bodies are worked out in total disregard of the fact that the formulae used are not meant for them and also, certain conditions have been applied to get the desired results.

CHAPTER 8

How Relations for Light Forced upon Other Bodies Goes Unnoticed in Relativity

The following analysis brings out how the relativistic relations, emerging out of the unique properties of light, have been hijacked for application on bodies/objects which have the corresponding properties nowhere near those used for the (derivation and validation of) relations.

The Contrived Adoption of γ as $1/\sqrt{1 - v^2/c^2}$, Which is Valid Only for Light:

The Lorentz transformation relations for distance are as follows, with a particular known value/expression for γ.

$$\left.\begin{array}{l}x' = \gamma(x - vt) \\ x = \gamma(x' + vt')\end{array}\right\} \ldots \ldots (1)$$

For a moment, however, **let us assume γ to be unknown**.

On separating t' and t, one-by-one, from the above two relations (1), one gets the following.

$$\left.\begin{array}{l}t' = \gamma\left[t - \left(1 - \frac{1}{\gamma^2}\right)\frac{x}{v}\right] \\ t = \gamma\left[t' + \left(1 - \frac{1}{\gamma^2}\right)\frac{x'}{v}\right]\end{array}\right\} \ldots \ldots (2)$$

To separate γ, we may multiply together both the relations of either (1) or (2), and write as follows.

$$\gamma = \left(1 - \frac{vt}{x}\right)\left(1 + \frac{vt'}{x'}\right)$$
$$\gamma = \left[1 - \left(1 - \frac{1}{\gamma^2}\right)\frac{x}{vt}\right]\left[1 + \left(1 - \frac{1}{\gamma^2}\right)\frac{x'}{vt'}\right] \quad \ldots \ldots (3)$$

Note: The first of relations (3) should not be used in cases where either $x = 0$ or $x' = 0$. Because in such cases, x or x', as the case may be, disappears from both of the relations (1), and the requirement becomes simply $\gamma = t'/t$ or $\gamma = t/t'$ respectively, as can be worked out from relations (1) itself. Similarly, the second of relations (3) should not be used in cases where either $t = 0$ or $t' = 0$ because in such cases, t or t', as the case may be, disappears from both of the relations (2), and the requirement becomes simply $\gamma = x'/x$ or $\gamma = x/x'$ respectively, as can be worked out from relations (2) itself.

It may be noted that irrespective of the value of γ, all substitutions of expressions from the above relations (1) and (2) in any one of them, or in (3), would always satisfy the relation, with γ disappearing.

Therefore, all values of γ conform to relations (1), (2) and (3).

However, for a given set of (x, t and v), all values of γ, except one, would result into the ratio x'/t' being different from x/t. On the other hand, the value of γ which ensures identity of the two ratios is as follows, and it is obvious from the first of relation (3).

$$\gamma = \frac{1}{\sqrt{1 - \frac{v^2 t^2}{x^2}}} = \frac{1}{\sqrt{1 - \frac{v^2}{(x^2/t^2)}}}$$

Therefore, for an object moving with a uniform velocity $u = x/t$, if one selects a value of γ, out of an infinite number of compatible values, which is equal to $1/\sqrt{1 - v^2/u^2}$, the velocity of the object would be the same in all co-directionally moving inertial frames. However, it would not conform to the Lorentz Transformation Condition, which is as follows.

$$c^2 t'^2 - x'^2 = c^2 t^2 - x^2$$

One the other hand, if the object was a light signal, it is only the value of γ, out of an infinite number of compatible values, which is equal to $1/\sqrt{1 - v^2/c^2}$, that would ensure constancy of its speed in all co-directionally moving inertial frames, and also conformity to the Lorentz Transformation Condition which was meant only for light, as shown in the following para.

On substitution of the expressions of x' and t' from (1) and (2) on LHS of the condition, we get

$$c^2 t'^2 - x'^2 = c^2 \left[\gamma \left[t - \left(1 - \frac{1}{\gamma^2}\right) \frac{x}{v} \right] \right]^2 - (\gamma(x - vt))^2$$

$$= \gamma^2 \left[c^2 t^2 + \left(1 - \frac{1}{\gamma^2}\right)^2 \frac{c^2 x^2}{v^2} - 2\left(1 - \frac{1}{\gamma^2}\right) \frac{c^2 xt}{v} - x^2 - v^2 t^2 + 2xvt \right]$$

$$= \gamma^2 \left[(c^2 t^2 - x^2) + \left(1 - \frac{1}{\gamma^2}\right)^2 \frac{c^2 x^2}{v^2} - 2\left(1 - \frac{1}{\gamma^2}\right) \frac{c^2 xt}{v} - v^2 t^2 + 2xvt \right]$$

$$= \gamma^2 \left[(c^2 t^2 - x^2) - \frac{v^2}{c^2} \left(c^2 t^2 - \left(1 - \frac{1}{\gamma^2}\right)^2 \frac{c^4}{v^4} x^2 \right) + 2xvt \left(1 - \left(1 - \frac{1}{\gamma^2}\right) \frac{c^2}{v^2}\right) \right]$$

A close examination of the expression reveals that for the following expression of γ,

$$1 - \frac{1}{\gamma^2} = \frac{v^2}{c^2} \quad \ldots \ldots (4)$$

the entire expression reduces to the following

$$c^2 t'^2 - x'^2 = c^2 t^2 - x^2$$

which is the objective.

It is reiterated that **the relation (4) is a condition which is true only for events of light**, according to (3), and it does not provide any proof of being true for other events which are saddled with $x \neq ct$ and $x' \neq ct'$.

In other words, it is only the events of light that conform to the Lorentz Transformation Condition, and that is because the light is the only known object that maintains constancy of its speed in all co-directionally moving inertial frames, which thereby, resulted into the value of γ as $1/\sqrt{1 - v^2/c^2}$.

Thus the contrived adoption of this value of γ, out of the infinite number of compatible values, for events other than those of light goes unnoticed, as all the requirements from the transformation relations to the Lorentz Transformation condition are met with this value of γ, though its own requirement in the form of $x = ct$ and $x' = ct'$ is not met by such events.

Consequently, the more is the event's distance-time ratio different from the speed of light in vacuum, the more is the discordance between the true and the worked out values in Relativity.

CHAPTER 9

Misconceptions about Time Dilation and Length Contraction

Abstract:

The common perception of the theory of Relativity is that moving bodies suffer from dilation of their time and contraction of length, and as a result of time dilation, their processes slowdown and their lifespans increase. There are innumerable particles around us, which suffer from these distortions for us, but we don't, although we are moving too, with respect to them, with all speeds and in all directions.

Does it not mean that we have positioned ourselves, going against the very theory of Relativity, as the privileged stationary frame where no such distortions occur? This question needs to be answered first, before casting our assumptions on the particles we claim to analyse.

The article discusses as to how these occurrences, emerging in specific contexts of Relativity, have incorrectly taken the form of a compulsory attribute of moving bodies/particles.

Introduction:

Time dilation and Length Contraction are the two terms that came to be widely used in modern physics. Numerous attempts – supposedly successful – have been made to explain many physical phenomena, and also to conceptualise/support many theories in modern physics.

The routine usage of the terms has, however, led to a disconnect from their parent theory, and a lot many practitioners of physics are found to be harbouring misconceptions about these occurrences.

The Two Together Make All Speeds Invariant!

Time dilation means more time in the stationary frame, or less time in the moving frame. Similarly, length contraction also means less length in the moving frame. The two together are in the right direction, from the viewpoint of conformity to Lorentz Transformation Condition, and any view of these being of opposite signs would contradict the Lorentz Transformation Condition. However, the prevailing perception, of both happening by the same factor $\sqrt{1 - \frac{v^2}{c^2}}$ in the moving frame, has grave consequences which are not taken note of. If the distance as well as the time in the moving frame change by the same factor, their ratio would be free from any changes, and therefore, all speeds of the stationary frame would remain unchanged in the moving frame. Thus a property, solely reserved for light, gets applicable to all moving bodies. So, nothing could be more obviously wrong than this. Even then, unfortunately, the perception has been ruling innumerable minds.

Distort for Whom?

If the current perception professes that the time dilation and length contraction happen within the moving bodies themselves, a natural question arises as to which benchmark these arise for. The universal impression is that these happen with respect to Earth **by default**, without realising that these are only observed from the Earth's frame, and do not actually happen within the bodies. Because a body is moving with respect to not only Earth but also other innumerable bodies. Thus it would have infinite number of velocities, with respect to the infinite number of particles around us, and all pairs are equivalent from Relativity point of view.

Does a moving body distort in space and time in infinite ways for the infinite number of benchmarks available? The question leads us to the correct answer i.e. no. The distortion is always one, with respect to only the observer judging it, and other benchmarks (or relatively moving bodies) do not matter. So, we reach the following conclusion.

Time dilation and length contraction are only the effects of Relativity, observed from other frames, and do not occur within the frame holding the time/clock or the length.

Under the conditions, it would be interesting, as well as educating, to recall the parent theory that gave birth to these phenomena, and more importantly, to examine under what conditions and to what extent these arose.

Parent Relations:

All would agree that the interpretations arose from the theory of Special Relativity, and their value is worked out from the Lorentz transformation given below.

$$\left.\begin{array}{l} x' = \dfrac{x - vt}{\sqrt{1 - \dfrac{v^2}{c^2}}} \\[2ex] t' = \dfrac{t - \dfrac{v}{c^2}x}{\sqrt{1 - \dfrac{v^2}{c^2}}} \end{array}\right\} \quad \ldots \ldots (1)$$

Here, all the terms are familiar. Remember that the non-primed frame (say, K) is stationary and the primed frame (say, K') is moving with a velocity of v with respect to the stationary frame, in the (+)ve x-direction. So, the non-primed parameters x and t are for the stationary frame, where these objects of observation i.e. x and t are placed. These objects are observed from the moving (primed) frame, with the corresponding results as x' and t' respectively.

This setup of relations has to be adhered to strictly under all conditions, for any correct interpretation.

For instance, when one has to consider the moving frame as stationary, and the stationary one as moving in the opposite direction, the same scheme has still to be followed. Now, x' and t' become the objects of observation, and x and t become the results. Therefore, the non-primed and the primed parameters have to swap their places, and v gets replaced by $-v$. With these changes, the above relations take the following form.

$$x = \frac{x' + vt'}{\sqrt{1 - \frac{v^2}{c^2}}}$$

$$t = \frac{t' + \frac{v}{c^2}x'}{\sqrt{1 - \frac{v^2}{c^2}}} \quad \bigg\} \quad \ldots \ldots (2)$$

Now, let there be two events, represented by (x_1, t_1) and (x_2, t_2) in stationary frame K. These are represented by (x'_1, t'_1) and (x'_2, t'_2) respectively in the moving frame K'.

Two relations similar to (1) can be written separately, between (x_1, t_1) and (x'_1, t'_1), and between (x_2, t_2) and (x'_2, t'_2). On taking difference of these two relations, and denoting the difference with Δ, the resultant relations take the following form.

$$\Delta x' = \frac{\Delta x - v\Delta t}{\sqrt{1 - \frac{v^2}{c^2}}}$$

$$\Delta t' = \frac{\Delta t - \frac{v}{c^2}\Delta x}{\sqrt{1 - \frac{v^2}{c^2}}} \quad \bigg\} \quad \ldots \ldots (3)$$

Similarly, the same exercise with relations (2) gives us the following relations.

$$\Delta x = \frac{\Delta x' + v\Delta t'}{\sqrt{1 - \frac{v^2}{c^2}}}$$

$$\Delta t = \frac{\Delta t' + \frac{v}{c^2}\Delta x'}{\sqrt{1 - \frac{v^2}{c^2}}} \quad \bigg\} \quad \ldots \ldots (4)$$

Now, the relations (3) and (4) are valid, in the two frames, for any length (Δx or $\Delta x'$) and the time interval at its ends (Δt or $\Delta t'$ respectively).

It must, however, be noted that a given length and its associated time interval are inseparably tied together in a frame, for the purpose of working out transformations in other frames.

Time Dilation:

The time dilation is stated to be happening in moving frames by a factor equal to $1/\sqrt{1-\frac{v^2}{c^2}}$. The origin of the moving frame is considered to be the same as the moving body. Thus the object of observation, i.e. the time interval, is in the moving frame, and we judge the results in Earth's (stationary) frame. As a result, the second of relations (4) becomes the appropriate one for this case.

Since the clock in the moving frame is constantly positioned at the frame's origin, the distance gap between any two ticks of the clock is zero i.e. $\Delta x' = 0$. On substituting it in the second of relations (4), one gets

$$\Delta t = \frac{\Delta t'}{\sqrt{1-\frac{v^2}{c^2}}}$$

This is the familiar relation for time dilation.

However, we must not lose sight of the fact that it is not the time ($\Delta t'$) between the two ticks of the moving clock that has changed but it is the time (Δt), as judged in the Earth's frame, which has become larger (by Lorentz Factor).

The rate of time for the moving clock always remains the same, and the moving clock, being stationary to itself, always gives the proper time to itself. **Thus no change can occur in the rate of processes, or lifespan, in moving particles/bodies themselves.**

Further, it may be interesting to note that the time dilation changes to time compression by the same factor, if the start and end of the time interval correspond to locations separated by a distance ($\Delta x'$) equal to $-v\Delta t'$. This also leads to $\Delta x = 0$ from the first of relations (4), meaning a situation where the two times correspond to the same location in the observer's (stationary) frame.

To summarise, we got time dilation with the condition $\Delta x' = 0$, and got time compression with the condition $\Delta x = 0$.

Thus the conclusion emerging out is as follows.

Time dilation is an effect of Relativity, with its value depending on the locations corresponding to time being observed, in frames other than the one holding the time, and it never occurs in the holding frame itself.

Length Contraction:

The moving objects are said to contract in length, in the direction of their motion, by a factor $\sqrt{1 - \frac{v^2}{c^2}}$. Let us examine this too, with respect to the parent relations.

Since the (object's) length is moving, and we judge the results in Earth's (stationary) frame, the first of relations (4) becomes the appropriate one for this case.

It may be recalled that in the previous case of time dilation, the co-parameter, distance, was kept the same at both the ends (ticks) of the time interval. Similarly, in this case too, if we keep the same value of the co-parameter, time, at both of its ends, we get similar results i.e. expansion.

That is, however, a rude shock, as we are **getting just the opposite of what we profess**.

So, to save the situation, we have to compulsorily resort to other favourable setups. A little application of mind would reveal that instead of choosing $\Delta t' = 0$, if we chose $\Delta t' = -\frac{v}{c^2} \Delta x'$, the desired result is obtained, as follows.

$$\Delta x = \Delta x' \sqrt{1 - \frac{v^2}{c^2}}$$

It may further be noted that the sought relation $\Delta t' = -\frac{v}{c^2} \Delta x'$ also leads to $\Delta t = 0$ from the second of relations (4), meaning the time corresponding to the two ends of the length should be the same in the observer's (stationary) frame.

To summarise, we got length elongation with the condition $\Delta t' = 0$, and length contraction with the condition $\Delta t = 0$.

It must again be noted here that these effects are noticed only in the other frame, and never occur in the frame where the observed length is lying.

So, the conclusion emerges as follows.

Length contraction is an effect of Relativity, with its value depending on the time corresponding to the ends of the length being observed, in frames other than the one holding the length, and it never occurs in the holding frame itself.

Conclusion:

It has been shown above from the parent relations, which serve as the basis for proclamations such as time dilation and length contraction, that these are nothing more than effects of Relativity, observed in other relatively moving frames, and do never occur in the frame holding the time/clock or the length.

Further, even the effects observed in other frames, do not always happen in the same ratio (by Lorentz factor or its inverse), but depend a lot on the values of the co-parameter selected at the two ends of the (length or time) interval.

CHAPTER 10

Commensurate Validations Required for Theory of Relativity

As all of us know, the theory of Relativity (Special) manifests itself in the form of Lorentz Transformation, and although the relations have been derived for a ray of light or photon, yet the same are used for transformation of all kinds of events i.e. all distance-time sets.

As a result, the backbone of the transformation relations i.e. the Lorentz Factor γ, though deriving itself naturally for events of light, becomes a contrived/imposed factor for other events. This forced adoption for the latter demolishes the integrity of the transformation relations for them, as discussed below.

The discussion below first shows as to how the value of γ as $1/\sqrt{1 - v^2/c^2}$ is unique to only the events of light, and any arbitrary value of it fits into the Lorentz transformation for other-than-light events. The following part (of the discussion) brings out as to what validation tests were necessary if we chose to forcibly adopt the current value of Lorentz Factor for the latter.

Let us take up the simplest case of an event in stationary frame (x, t) being observed by a another frame moving with a uniform velocity v in (+)ve x-direction, with respect to the stationary frame. The event as observed in the moving frame is denoted by (x', t').

The Lorentz transformation relations for distance are as follows, for all kinds of events (distance-time sets), with a particular known value/expression for γ.

$$\left.\begin{array}{l} x' = \gamma(x - vt) \\ x = \gamma(x' + vt') \end{array}\right\} \quad \ldots \ldots (1)$$

For a moment, however, **let us assume γ to be unknown.**

On separating t' and t, one-by-one, from the above two relations (1), one gets the following.

$$\left.\begin{array}{l} t' = \gamma\left[t - \left(1 - \frac{1}{\gamma^2}\right)\frac{x}{v}\right] \\ t = \gamma\left[t' + \left(1 - \frac{1}{\gamma^2}\right)\frac{x'}{v}\right] \end{array}\right\} \quad \ldots\ldots (2)$$

To separate γ, we may multiply together both the relations of either (1) or (2), and write as follows.

$$\left.\begin{array}{l} \frac{1}{\gamma^2} = \left(1 - \frac{vt}{x}\right)\left(1 + \frac{vt'}{x'}\right) \\ \frac{1}{\gamma^2} = \left[1 - \left(1 - \frac{1}{\gamma^2}\right)\frac{x}{vt}\right]\left[1 + \left(1 - \frac{1}{\gamma^2}\right)\frac{x'}{vt'}\right] \end{array}\right\} \quad \ldots\ldots (3)$$

Note: The first of relations (3) should not be used in cases where either $x = 0$ or $x' = 0$. Because in such cases, x or x', as the case may be, disappears from both of the relations (1), and the requirement becomes simply $\gamma = t'/t$ or $\gamma = t/t'$ respectively, as can be worked out from relations (1) itself. Similarly, the second of relations (3) should not be used in cases where either $t = 0$ or $t' = 0$ because in such cases, t or t', as the case may be, disappears from both of the relations (2), and the requirement becomes simply $\gamma = x'/x$ or $\gamma = x/x'$ respectively, as can be worked out from relations (2) itself.

Now, if we take up any arbitrary value of γ, and find out the values of x' and t', with a given set of x and t, from the first of relations (1) and (2), and substitute them back into the second of the same relations, we get back the values of x and t we started with. The same is true with interchange of the observers too. After knowing the parameters of both the frames/observers, if we substitute them in relation (3), we get back the value of γ we started with.

Thus all values of γ turn out to be conforming to the three relations i.e. (1), (2) and (3).

However, for a given set of ($x \neq 0$, $t \neq 0$ and v), all values of γ, except one, would result into the ratio x'/t' being different from x/t. Any value of γ requires a particular relation between the distance-time ratios in the two frames, which is as follows.

$$\frac{x'}{t'} = \frac{x - vt}{t - \left(1 - \frac{1}{\gamma^2}\right)\frac{x}{v}} = \frac{x}{t}\left[\frac{1 - \frac{vt}{x}}{1 - \left(1 - \frac{1}{\gamma^2}\right)\frac{x}{vt}}\right]$$

Consequently, the value of γ which is compatible with identity of the two ratios is as follows, and it is also obvious from the first of relation (3).

$$\gamma = \frac{1}{\sqrt{1 - \frac{v^2 t^2}{x^2}}} = \frac{1}{\sqrt{1 - \frac{v^2}{(x^2/t^2)}}}$$

Therefore, for an object moving with a uniform velocity $u = x/t$, if one selects a value of γ, out of an infinite number of compatible values, which is equal to $1/\sqrt{1 - v^2/u^2}$, the velocity of the object would be the same in all co-directionally moving inertial frames. However, it would not conform to the Lorentz Transformation Condition, which is as follows.

$$c^2 t'^2 - x'^2 = c^2 t^2 - x^2$$

One the other hand, for a light signal, it is only the value of γ, out of an infinite number of compatible values, which is equal to $1/\sqrt{1 - v^2/c^2}$, that would ensure constancy of its speed in all co-directionally moving inertial frames, and also conformity to the Lorentz Transformation Condition which was meant only for light, as shown in the following para.

On substitution of the expressions of x' and t' from (1) and (2) on LHS of the condition, we get

$$c^2 t'^2 - x'^2 = c^2 \left[\gamma\left[t - \left(1 - \frac{1}{\gamma^2}\right)\frac{x}{v}\right]\right]^2 - (\gamma(x - vt))^2$$

$$= \gamma^2 \left[c^2t^2 + \left(1 - \frac{1}{\gamma^2}\right)^2 \frac{c^2x^2}{v^2} - 2\left(1 - \frac{1}{\gamma^2}\right)\frac{c^2xt}{v} - x^2 - v^2t^2 + 2xvt \right]$$

$$= \gamma^2 \left[(c^2t^2 - x^2) + \left(1 - \frac{1}{\gamma^2}\right)^2 \frac{c^2x^2}{v^2} - 2\left(1 - \frac{1}{\gamma^2}\right)\frac{c^2xt}{v} - v^2t^2 + 2xvt \right]$$

$$= \gamma^2 \left[(c^2t^2 - x^2) - \frac{v^2}{c^2}\left(c^2t^2 - \left(1 - \frac{1}{\gamma^2}\right)^2 \frac{c^4}{v^4}x^2\right) + 2xvt\left(1 - \left(1 - \frac{1}{\gamma^2}\right)\frac{c^2}{v^2}\right) \right]$$

A close examination of the expression reveals that for the following expression of γ,

$$1 - \frac{1}{\gamma^2} = \frac{v^2}{c^2} \quad \ldots \ldots (4)$$

the entire expression reduces to the following

$$c^2t'^2 - x'^2 = c^2t^2 - x^2$$

which is the objective.

It is reiterated that **the relation (4) is a condition which is true only for events of light**, according to (3), and it cannot be true for other events which are saddled with $x \neq ct$ and $x' \neq ct'$.

In other words, it is only the events of light that conform to the Lorentz Transformation Condition, and that is because the light is the only known object that maintains constancy of its speed in all co-directionally moving inertial frames, which thereby, resulted into the value of γ as $1/\sqrt{1 - v^2/c^2}$.

It is added here that this expression of γ leads to the term $\left(1 - \frac{1}{\gamma^2}\right)\frac{x}{v}$ of relations (2) reducing to vx/c^2, which ensures conformity to Maxwell's EM wave equation in both the frames. This further confirms unique relevance of the Lorentz Factor only for light.

Thus the contrived adoption of this value of γ, out of the infinite number of compatible values, for events other than those of light goes unnoticed, as all the requirements from the transformation relations to the Lorentz Transformation condition are met with this value of γ, though its own requirement in the form of $x = ct$ and $x' = ct'$ is not met by such events.

Consequently, the more is the event's distance-time ratio different from the speed of light in vacuum, the more is the discordance between the value mandated by the Lorentz Transformation Condition, and that worked out by Lorentz transformation.

EXAMPLE:

Let us take up an example of a particle moving uniformly in x-direction with a relative velocity of $0.90c$, with respect to the stationary observer. There is another observer moving with a velocity of $0.89c$ in the same direction, with respect to the stationary observer. Both the observers record their own measurements of distance and time traversed by the particle. To reiterate, the stationary observer's parameters are non-primed (i.e. x and t) and those of the moving observer are primed (i.e. x' and t').

As already shown, all values of γ would conform to relations (1), (2) and (3). Further, unless one knows all the four values i.e. x, t, x' and t', one cannot fix a unique value of γ from relations (3). Therefore, to make it simple, we take up a value of γ, which arises when the ratios x/t and x'/t' become the same, meaning when the velocity (u) of the moving particle is the same in both the frames, and this is given by

$$\gamma = \frac{1}{\sqrt{1 - \frac{v^2 t^2}{x^2}}} = \frac{1}{\sqrt{1 - \frac{v^2}{u^2}}}$$

In the present case, with $v = 0.89c$ and $u = 0.90c$, $\gamma = 6.726916$.

This value of γ conforms to relations (1), (2) and (3), but does not conform to the Lorentz Transformation condition, as already shown in the previous section.

On the other hand, when one takes the case of a light signal in place of the moving particle, $u = c$, and the corresponding value of γ becomes 2.193172, which conforms to the Lorentz Transformation Condition.

Presently, as a general rule, we take γ as 2.193172, in spite of its worked out value being 6.726916. The gap is too wide to be neglected.

What emerges out here is that the value of γ works out in such cases as $1/\sqrt{1 - v^2/u^2}$, but it continues to be forced upon as $1/\sqrt{1 - v^2/c^2}$ to achieve conformity to the Lorentz Transformation Condition.

Further, the impact of ratio (x/t) being different from c is not as large on the ratio (x'/t') as it is on the individual values of x' and t'. This is because there is no direct impact of the term γ on the ratio, as it is in case of x' and t' individually.

For example, in the present case, the two values of γ, different by 3.07 times, would coincide when the ratio (x'/t') was only $(1/0.9)$ times (x/t). Going a step further, if the velocity of the particle u was $0.891c$, the value of γ would jump to 21.1128, making it 9.63 times, and then the ratio (x'/t') will have to be only $(1/0.891)$ times (x/t), for this value of γ to reduce to the current value of the Lorentz Factor.

Forced Adoption of Lorentz Factor for Other-than-Light Events Requires Validation Tests:

Despite the above discussion, one may still argue that the Lorentz Factor $1/\sqrt{1 - v^2/c^2}$, though deriving itself inherently for the events of light, is also suitable for transformation of other events, as it is the only value of γ which conforms to the Lorentz Transformation condition, a mandatory requirement for all transformations, and it was also established by numerous tests and observations (though truncated).

However, as shown in the previous section, adoption of this value of γ requires the following condition to be satisfied.

$$\frac{x'}{t'} = \frac{x - vt}{t - \left(1 - \frac{1}{\gamma^2}\right)\frac{x}{v}} = \frac{x}{t}\left[\frac{1 - \frac{vt}{x}}{1 - \frac{vx}{c^2 t}}\right]$$

To simply it further, if the distance-time ratio of the event in stationary frame is expressed as αc i.e. $x/t = \alpha c$, the requirement can be expressed as follows, with $\beta = v/c$.

$$\frac{x'}{t'} = \frac{x}{t}\left[\frac{1 - \beta/\alpha}{1 - \beta\alpha}\right]$$

For events of light, the relation is always satisfied, as $\alpha = 1$.

However, whether the requirement, in its full form (non-zero values of distance and time together), is met by other events still remains to be tested, as it requires a frame moving at relativistic speed and also capable of measuring non-zero values of both - distance as well as time - of events.

It may be recalled that the observations so far, of length contraction or time dilation, are based on freezing either the time or location respectively between two events, in the observed/observing frame. In such controlled situations, the requirement of γ collapses to a ratio of mere distances or times in the two frames, as explained in Notes under relation (3).

For example, with $\beta = 0.9$ and $\alpha = 0.1$, the full requirement is $x'/t' = -8.79121(x/t)$ or $x'/x = -8.79121(t'/t)$. Further, by only increasing the value of α to 0.89, the requirement changes to $x'/t' = -0.05646(x/t)$ or $x'/x = -0.05646(t'/t)$. However, the current tests of Special Relativity are limited only to showing $t' = \gamma t$. In the instant example, with $\gamma = 1/\sqrt{1 - (0.9)^2} = 2.2942$, the tests would merely establish $t' = 2.2942t$, which is too different from the full requirement in the two cases.

Therefore, the current tests hardly offer any justification on the compatibility of the Lorentz Factor in the transformation relations for events other than light. The true test can be applied only by recording together non-zero values of both – distance as well as time – in both the frames.

Very importantly, while waiting for results of such full confirmatory tests, the second postulate of Special Relativity needs to be suitably changed/supplemented to adopt the Lorentz Factor for other-than-light events, so as to grant sanctity to it, as the current postulate is good for only events of light, and the Lorentz Factor does not emerge out naturally in the process of derivation, but has been imposed, for events other than those of light. The same would get institutionalized if the full tests confirm the assumption.

CHAPTER 11

Part Application of Relativity Lead to Errors & Paradox in Doppler Effect and Incorrect Explanation for Muons Reaching Earth

Special Relativity envisages neither time dilation nor length contraction in isolation, independent of each other; but instead, the two occur simultaneously, in any transformation, with values that together have to mandatorily conform to the Lorentz Transformation Condition. However, if one of them is taken as zero to start with theoretically, the transformation obviously reduces to the other parameter (distance or time) with the non-zero value. Even though such assumed cases are nearly impossible in practice, their results are orchestrated as general laws of nature that exist independently, without any riders. The oft-quoted phenomena like 'time ticking slower' and 'length contraction' in moving frames are examples. These are quoted as if these happen always, without any conditions, which is incorrect. The reason would be clearer in the discussions ahead.

To render the classical Doppler Effect with Relativity, time dilation is applied to the time period of the electromagnetic waves in the receiver's frame, ignoring the inseparable length contraction of its wavelength. Further, since Relativity does not discriminate, in respect of relative motion, between the source and the receiver, one could consider the source as moving to apply the time dilation on the emitted wave. In such a case, the universally professed redshift change to blueshift, thus creating a paradox.

Similarly, in case of muons, the time dilation is applied to its mean lifespan to explain as to how these, with an extremely short lifespan of about 2.2 microseconds, are able to cover disproportionately large distances in Earth's atmosphere. Here again, no heed is paid to the inseparable length transformation occurring in almost the same proportion. Explanations are also presented from the perspective of muons experiencing drastic length contraction of Earth's atmosphere, while ignoring the inseparable contraction of all the timespans in Earth's frame. When Relativity is applied in entirety, the current explanations fail to sustain in face of the mandatory Lorentz Transformation Condition.

Introduction -The Mandatory Requirement:

The transformations of time and length are inseparable in Relativity, and one cannot occur without the other. This mandatory requirement is stipulated by the Lorentz Transformation Condition as follows.

$$c^2 t'^2 - x'^2 = c^2 t^2 - x^2 = Constant$$

where one of the frames – non-primed or primed – is stationary and the other one is moving with respect to the other with a uniform velocity along x-axis.

On rearranging, the Condition becomes

$$c^2 t'^2 - c^2 t^2 = x'^2 - x^2 \ldots \ldots (1)$$

The Condition stipulates that, on transformation from one frame to the other, if time of an event changes in a particular direction, (+)ve or (-)ve, the corresponding distance of the event has also to change in the same direction. The relation between the magnitudes of their changes is also decided by the Condition.

That leads to the conclusion that if time of a frame appears increased in another frame, all the lengths of the former would necessarily have to appear longer in almost (not exactly) the same proportion in the latter. If, however, one is dealing with transformation of distance and time of an electromagnetic signal/ray, the ratio of the times, in the two frames, has to be exactly equal to that of distances, as $x = ct$ and $x' = ct'$.

The two cases are discussed below.

I. Relativistic Doppler Effect:

Let us take the case of the source (S) of an electromagnetic wave and its receiver (R) moving away from each other with a relative velocity v, along the line joining them.

Let the frequencies of the wave in their frames respectively be f_s And f_r.

The classical Doppler Effect is expressed by the following relation

$$f_r = \frac{f_s}{1+\beta}$$

where $\beta = \frac{v}{c}$

Now, the current practice to render the classical effect with Relativity is to apply time dilation to the time period of the wave in the moving frame.

If the receiver is considered moving, the time period of the received wave has to be multiplied by a factor of $\sqrt{1-\beta^2}$ and therefore, the frequency f_r to be multiplied by a factor of $1/\sqrt{1-\beta^2}$. On doing so and rearranging, one gets

$$\frac{f_s}{f_r} = \sqrt{\frac{1+\beta}{1-\beta}}$$

Thus one gets the formula extensively used in astronomy, which predicts redshift observed by receivers on the Earth.

However, according to Relativity, the source could also be considered as moving, with the receiver considered as stationary. When one takes to this option, the time period of the emitted wave is to be multiplied by a factor of $\sqrt{1-\beta^2}$ and therefore, the frequency f_s to be multiplied by a factor of $1/\sqrt{1-\beta^2}$. On doing so, one gets the following relation

$$\frac{f_s}{f_r} = (1+\beta)\sqrt{1-\beta^2}$$

The result predicts a redshift for $\beta < 0.6185$, no shift for $\beta = 0.6185$ and a blueshift for $\beta > 0.6185$.

This is widely different from what is achieved by taking to the previous option and thus, it leads to a paradox.

It highlights the pitfalls of motivated application of only parts of Relativity to physical phenomena, in disregard of the Condition at (1) above.

Application in Entirety Leaves no Room for Paradox but Changes Predictions:

Application of Relativity in its entirety, in fact, dispenses altogether with working out of the classical Doppler Effect. Further, the Lorentz transformation of a single electromagnetic wavelength and its time period in either of the frames lead to the same result, which is shown below.

Let λ_s, t_s and f_s be the wavelength, the time period and the frequency of the wave emitted from the source. Similarly, let the corresponding parameters of the wave received by the receiver be λ_r, t_r and f_r respectively.

Imp: Before proceeding further, it is reminded that the structure of the Lorentz transformation requires that the stationary frame holds the distance and time parameters *to-be-transformed* (connected by expressions on RHS), and the moving frame holds the corresponding *transformed* parameters (appearing as sole parameters on LHS).

The two equivalent setups, based on the principle of reciprocity of velocity stipulated by Relativity, are taken below, to work out the change in frequency of emitted/received wave on account of the relative motion between the source and the observer.

A. Source Stationary and Receiver Moving:

Let an electromagnetic wave crest incoming from the source (stationary frame) coincide with the origin of the moving frame i.e. receiver, at the instant the latter starts moving with a velocity v in the direction of motion of the wave. The assumption is in accordance with the setup of increasing distance between the source and the receiver.

The origins of time and distance of the stationary frame (source) are also considered coinciding with those of the moving frame (receiver) at start.

In the stationary frame (source), after a time equal to t_s, the wave would have moved by a distance equal to λ_s towards the receiver and the receiver would have moved by a distance vt_s in the same direction, from the origin.

On applying the Lorentz transformation on a single wavelength and its time period in the stationary frame (source) to get their values in the moving frame (receiver), one gets following relations

$$\left.\begin{array}{l} \lambda_r = \gamma(\lambda_s - vt_s) \\ t_r = \gamma\left(t_s - \dfrac{v\lambda_s}{c^2}\right) \end{array}\right\}$$

Or,

$$\left.\begin{array}{l} \lambda_r = \dfrac{1}{\sqrt{1-\beta^2}}\left(\lambda_s - \dfrac{v}{c}\lambda_s\right) \\ t_r = \dfrac{1}{\sqrt{1-\beta^2}}\left(t_s - \dfrac{v}{c}t_s\right) \end{array}\right\}$$

Or,

$$\left.\begin{array}{l} \lambda_r = \sqrt{\dfrac{1-\beta}{1+\beta}}\,\lambda_s \\ t_r = \sqrt{\dfrac{1-\beta}{1+\beta}}\,t_s \end{array}\right\}$$

The relations obtained represent blueshift in the receiver's frame, which is just the inverse of what is currently being worked out by applying only the time dilation in the receiver's frame.

It is, however, noteworthy that the soundness of the results can be verified by dividing the first relation by the second, as follows.

$$\frac{\lambda_r}{t_r} = \frac{\lambda_s}{t_s} = c$$

which is in accordance with the second postulate of Relativity i.e. constancy of light speed in all inertial frames. The speed of the electromagnetic wave is the same at the receiver as at the source.

B. Receiver Stationary and Source Moving:

At the origin of the moving frame (source), let a wave crest be emitted towards the receiver, at the instant the former starts moving with a velocity v in the direction opposite to that of the emitted wave. The assumption is once again in accordance with the setup of increasing distance between the source and the receiver.

The origins of time and distance of the stationary frame (receiver) are also considered coinciding with those of the moving frame (source) at start.

In the stationary frame (receiver), after a time equal to t_r, the wave would have moved by a distance equal to λ_r towards the receiver and the source would have moved by a distance vt_r in opposite direction, from the origin.

On applying the Lorentz transformation on a single wavelength and its time period in the stationary frame (receiver) to get their values in the moving frame (source), one gets following relations, with v getting replaced by $-v$ in view of the source moving opposite to the direction of the wave.

$$\left. \begin{array}{l} \lambda_s = \gamma(\lambda_r + vt_r) \\ t_s = \gamma\left(t_r + \dfrac{v\lambda_r}{c^2}\right) \end{array} \right\}$$

Or,

$$\left. \begin{array}{l} \lambda_s = \dfrac{1}{\sqrt{1-\beta^2}}\left(\lambda_r + \dfrac{v}{c}\lambda_r\right) \\ t_s = \dfrac{1}{\sqrt{1-\beta^2}}\left(t_r + \dfrac{v}{c}t_r\right) \end{array} \right\}$$

Or,

$$\left.\begin{aligned} \lambda_s &= \sqrt{\frac{1+\beta}{1-\beta}}\,\lambda_r \\ t_s &= \sqrt{\frac{1+\beta}{1-\beta}}\,t_r \end{aligned}\right\}$$

Or,

$$\left.\begin{aligned} \lambda_r &= \sqrt{\frac{1-\beta}{1+\beta}}\,\lambda_s \\ t_r &= \sqrt{\frac{1-\beta}{1+\beta}}\,t_s \end{aligned}\right\}$$

which is the same as obtained previously in sub-section 'A'.

Thus the results are the same, no matter which one of the two – the source or the receiver - is considered moving. This leaves no room for any paradox.

It is also conceivable that **there cannot be any classical Doppler Effect**, in view of the speed of light being the same with respect to the source as well as to the receiver.

The above results show that when Relativity is applied in its entirety, the Relativistic Doppler Effect predicts a blueshift, on the earth, of the electromagnetic waves received from receding stars and galaxies. This is just the opposite of what is professed. The phenomena of redshift on the Earth, therefore, requires explanations from factors other than Relativity, and calls for identification of agents leading to loss of energy of the incoming photons.

II. The Enigma of Muons Reaching Earth:

The enigma is: a muon, with a mean lifespan of 2.2 microseconds and travelling with a speed as high as $0.999c$ can travel only a distance of $0.999 \times 300000 \times 2.2 \times$

10^{-6} = 0.659 km = 659 m before decaying. With this magnitude of travel capacity, even those generated at the heights of 15 km are detected on the Earth's surface.

Such phenomena are explained by invoking Relativity in part i.e. either time dilation or length contraction, in contradiction with the mandatory requirements of Relativity highlighted in the opening section titled "Introduction -The Mandatory Requirement".

The **current explanations** from both the approaches are first discussed below, followed by the exercise with complete (not part) application of Relativity.

A. Current Explanation With Time Dilation approach:

Muons live for 2.2 microseconds of their time, which observers on the Earth would measure $2.2/\sqrt{1 - v^2/c^2}$ microseconds. With the figures assumed above, this works out to 49.2 microseconds, which translates into a travel distance of $0.999 \times 300000 \times 49.2 \times 10^{-6}$ = 14.75 km. This is quite close to 15 km. Thus it makes arrival of muons possible on the Earth.

B. Current Explanation With Length Contraction approach:

For the muon, in its own frame, the entire length of Earth's atmosphere, say 15 km, is treated as moving upwards, and therefore, it appears to it as contracted by the factor i.e. $\sqrt{1 - v^2/c^2}$. With the figures assumed above, this leads to a contracted length of only 0.671 km or 671 m, which is easily covered by the muon within its lifespan of 2.2 microseconds.

Correction:

The above explanations fail to sustain, when **corrected below** by applying together both the inseparable effects of time dilation and length contraction.

Invoking the Condition at (1) above, let the primed parameters be of the Earth and the non-primed parameters be of a muon.

A1. Time Dilation Approach Supplemented with Length Transformation:

The Condition with relation (1) may be rearranged as follows.

$$x'^2 = c^2 t'^2 - c^2 t^2 + x^2$$

On substituting the expression of t' from sub-section 'A' above, one gets

$$x'^2 = c^2 \left(\frac{1}{1-\beta^2} - 1\right) t^2 + x^2 = c^2 t^2 \frac{\beta^2}{1-\beta^2} + x^2$$

Therefore, the expression for length in the Earth's frame becomes as follows.

$$x' = \sqrt{\frac{\beta^2}{1-\beta^2} c^2 t^2 + x^2}$$

On changing the units of c in a more convenient form, we have $c = 300$ m/microsec.

On substituting $t = 2.2$ microseconds and $\beta = 0.999$, as taken in sub-section 'A' above, the expression for x' becomes as follows, with unit of distances in meters.

$$x' = \sqrt{\frac{(0.999 \times 300 \times 2.2)^2}{1-(0.999)^2} + x^2} = \sqrt{217473354 + x^2} \text{ meters}$$

If x is taken as the muon's proper length in meters, the term x^2 may be neglected in the above expression, and its length in the Earth frame, x' works out to $14747\ m \approx 14.8\ km$.

It shows that a negligible length of muon should also be seen, in the Earth's frame, as lengthened to ≈ 14.8 km, along with the lengthening of its lifespan to 49.2 microseconds.

This obviously is impossible and thus, Relativity fails to explain the phenomena.

B1. Length Contraction Approach Supplemented with Time Transformation:

The Condition with relation (1) may be rearranged as follows.

$$c^2 t^2 = c^2 t'^2 - x'^2 + x^2$$

On substituting the expression of x from sub-section 'B' above, one gets

$$t = \sqrt{t'^2 + \frac{-x'^2 + x'^2(1-\beta^2)}{c^2}} = \sqrt{t'^2 - \beta^2 \frac{x'^2}{c^2}}$$

On changing the units of c in a more convenient form, we have $c = 0.3$ km/microsec.

On substituting $x' = 15$ km and $\beta = 0.999$, as taken in sub-section 'B' above, the expression for t becomes as follows, with unit of time in microseconds.

$$t = \sqrt{t'^2 - \left(\frac{0.999 \times 15}{0.3}\right)^2} = \sqrt{t'^2 - 2495} \text{ microseconds} \quad \ldots\ldots\ldots\ldots\ldots\ldots\ldots\ldots\ldots(1)$$

From the above, it is obvious that $t = 0$ for $t'^2 = 2495$

Or, $t = 0$ for $t' = \sqrt{2495} = 49.95$ microseconds.

It means that for a time span of 49.95 microseconds in the Earth's frame, the corresponding time in the muon's frame is zero, and for all lesser timespans, the corresponding time in the muon's frame become imaginary.

Its implications may be understood as follows.

Let us first assume that numerous stationary or slow moving muons are also getting generated continuously on the Earth's surface (say in a laboratory), in addition to those in the atmosphere. The muons generated in the Earth's atmosphere at a height of 15 km are able to reach the surface because the 15 km length contracts for them to 0.671 km, as the atmosphere is treated as moving upwards with respect to the muon with the same velocity. The atmospheric muons would be watching this upward motion for their proper lifespan of 2.2 microseconds.

Quite obviously, at the instant when the Earth's surface and the atmospheric muons meet, all the muons generated in the laboratory in the last 2.2 microseconds (Earth's time) should be available to meet with the atmospheric muons.

From the atmospheric muon's reference frame, while the atmosphere's length of 15 km reduces to 0.671 km for them, even the lifespans of the muons generated on the Earth in the last 2.2 microseconds (Earth's time) would get transformed, in accordance with the Condition stated by (1). However, as shown above, the transformation of the lifespans would lead to imaginary values, as the lifespans of 2.2 microseconds are too less than the threshold value of 49.95 microseconds, as calculated above.

Thus, the explanation offered by length contraction of atmosphere also fails to sustain in face of the Condition.

Conclusion:

The part applications of Relativity, which may well be termed as a motivated steps, pervade everywhere in physics. The parts like time dilation and length contraction are incorrectly treated as independent and separate phenomena under Relativity, mutually exclusive of each other.

The Relativity does not envisage any such separation, but physicists have been applying it only by parts to force-explain numerous phenomena of great importance.

It has been shown in the article that the part application, by way of only time dilation on Doppler Effect, has been projecting redshift from receding stars and galaxies, but the full application projects blueshift. The part application, though incorrect, gratifies us, as the projections somewhat go in line with the observations. The misplaced complacency, however, holds us from exploring other factors/theories that go behind so much of mystery still prevailing in the domain of cosmology.

The later part of the paper has also shown that the explanations of extremely short-lived muons reaching the Earth's surface after crossing a 15 km long atmosphere, by part application of Relativity (either time dilation or length contraction), fail to stand up to the mandatory requirements of the Lorentz Transformation Condition. Such incorrect explanations have been holding us, similar to the previous case, from search of other explanations from quantum theory or a new particle theory.

CHAPTER 12

How Einstein Interchanged the Object and Result of Transformation to Show Length Contraction and Time Dilation in Moving Frames

Introduction:

In Relativity, one may consider either of the two frames as stationary and other one as moving. However, one has to maintain the discipline that the objects of transformation are placed in the stationary frame, which are used as the benchmarks, with respect to which the results observed by the moving frame are worked out and compared. This is also a necessary requirement of the Lorentz transformation relations which are derived with this very structure.

It is, however, commonly noticed that while working out transformations of distance and time from the relativity relations and drawing conclusions therefrom, the objects of transformation and the results of transformation are mistakenly or purposefully interchanged. This obviously leads to incorrect conclusions spreading confusion among the practitioners of relativity.

Such a mess up has been done by Einstein himself in his 1916 book titled "Relativity: The Special and The General Theory", Chapter 12 (The Behaviour of Measuring-Rods and Clocks in Motion), to show length contraction and time dilation in moving objects.

The common sense, however, dictates that no transformation of length or time on a moving object can occur, similar to stationary ones, as the length and the time within any frame are the invariant proper values, independent of the object's state of motion.

It is the judgement of, or the measurement from, the other frame (which is relatively moving with respect to the former) that shows transformation.

The articles discusses these discrepancies in the aforesaid book.

Remember:

It may be recalled that the transformation relations are based on the assumption that x and t of the stationary frame are the objects of transformation and x' and t' are the corresponding results when observed from the frame moving in (+)ve x-direction. So, application of the relations would be correct only if a correct bifurcation is made between what is being measured and what represents the results.

Further, it must be ensured that the frame carrying the object of transformation is designated as stationary and the one observing the results is treated as moving, no matter what the given setup is. Consequently, the transformation results i.e. contraction, elongation/expansion, dilation etc. should be referred from the moving frame, and with respect to the stationary frame.

Examples:

Two such examples are presented below from the aforesaid book.

EXAMPLE 1

We will see here how the objects of transformation (as per the relations) have been shown as results, and the results (as per the relations) are treated as objects of transformation.

The above chapter, showing length contraction, begins with the following.

> Place a metre-rod in the x'-axis of K' in such a manner that one end (the beginning) coincides with the point x'=0 whilst the other end (the end of the rod) coincides with the point x'=1. What is the length of the metre-rod relatively to the system K? In order to learn this, we need only ask where the beginning of the rod and the end of the rod lie with respect to K at a particular time t of the system K. By means of the first equation of the Lorentz transformation the values of these two points at the time t = 0 can be shown to be

$$x_{(beginning\ of\ rod)} = 0\sqrt{1 - \frac{v^2}{c^2}}$$

$$x_{(end\ of\ rod)} = 1\sqrt{1 - \frac{v^2}{c^2}}$$

the distance between the points being

$$\sqrt{1 - \frac{v^2}{c^2}}$$

But the metre-rod is moving with the velocity v relative to K. It therefore follows that the length of a rigid metre-rod moving in the direction of its length with a velocity v is $\sqrt{1 - v^2/c^2}$ *of a metre.*

OBSERVATIONS

1. First of all, the Lorentz transformation derived for propagation of a light signal is being used for a fixed metre-rod, which is unacceptable. Further, at t = 0 or at t′ = 0, the propagating light cannot be at both the ends of the rod. At t′ = 0, it is already assumed at start of the derivation that x = 0, x′ = 0 and t = 0; so, all terms are zero and no result can come out of it.

2. If, however, we choose to procced further overlooking the above objections, we are faced with another error and that is - the object of transformation and the result of transformation have been interchanged. The same is explained below.

   It may be recalled that in the derivation of transformation equations (eq.8), x and t are the objects of transformation and the results of transformation are respectively x′ and t′. The relation has correctly been used to reverse calculate the values of object x in frame K, which would correspond to the given values of x′ (i.e. 0 and 1) in the moving frame K'. The results are known here and the measurements of objects are to be found out.

Now, let us examine the concluding statement in the last line "the length of a rigid metre-rod moving in the direction of its length with a velocity v is $\sqrt{1 - v^2/c^2}$ of a metre." It means the moving meter-rod is the object of transformation with known measurements i.e. 1 and its length $\sqrt{1 - v^2/c^2}$ is the result of transformation as noted by a stationary observer. This is contrary to the designations of object and the result in the relation used, which are $\sqrt{1 - v^2/c^2}$ and 1 respectively.

Thus the statement should correctly be as follows.

A rod of length $\sqrt{1 - \frac{v^2}{c^2}}$ lying in the stationary frame K is measured, at any given time of frame K, as being of length 1 from the moving frame K'.

Further, by relativity, the rod in frame K may also be considered moving (in opposite direction) while treating frame K' as stationary. So, the above statement may also be rephrased as follows.

A moving rod (of stationary length $\sqrt{1 - \frac{v^2}{c^2}}$) measures elongated (to a length 1) when judged by a stationary observer in a snapshot.

Further, the condition of 'a given time/ in a snapshot' in frame K' is very important, as analysed in detail in chapter 6.

This highlights the need for correct assignment of the object of transformation (x, t) and result of transformation (x', t') while using the transformation relations.

EXAMPLE 2

Referring to the same chapter as for example 1, towards the end, it is mentioned.......

Let us now consider a seconds-clock which is permanently situated at the origin (x'=0) of K'. t'=0 and t'=1 are two successive ticks of this clock. The first and fourth equations of the Lorentz transformation give for these two ticks:

$$t = 0$$

and

$$t = \frac{1}{\sqrt{1 - \frac{v^2}{c^2}}}$$

As judged from K, the clock is moving with the velocity v; as judged from this reference-body, the time which elapses between two strokes of the clock is not one second, but

$$\frac{1}{\sqrt{1 - \frac{v^2}{c^2}}}$$

seconds, i.e. a somewhat larger time. As a consequence of its motion the clock goes more slowly than when at rest. Here also the velocity c plays the part of an unattainable limiting velocity.

OBSERVATIONS

1. Here again, the Lorentz transformation derived for the time elapsed in propagation of a light signal is being used for a seconds clock fixed at $x' = 0$. Thus there is no compatibility in the assumptions of derivation and its application; therefore, this too is unacceptable. Further, at $x' = 0$, the propagating light cannot be at both the clicks of the clock.

2. In the context of the mistake pointed out at sr. no. (2) of the last observations on the first example, it may be noted that while applying the transformation relation of time here, frame K has correctly been treated as the moving frame and frame K' as stationary in which the clock is fixed.

However, the object and the result of transformation have been interchanged. Therefore, the statement "As a consequence of its motion the clock goes more slowly than when at rest" requires recast, as the clock is at rest in frame K' having its time as the proper time (which is also the same in stationary state). It is only the measurement from frame K that is different from the proper time. Consequently, the correct statement should be "As a consequence of its motion, the moving clock is judged, by a stationary observer, as running faster than when it is at rest (in the clock's own stationary frame)".

Further, it may be noted that at $t' = 1$, although $x' = 0$ yet x is non-zero i.e. $x = 1c$ in accordance with the derivation.

Further, the condition of freezing the location of the clock is noteworthy. Its implications have been discussed in detail in chapter 6.

COMMON OBSERVATION

The results of length contraction and time dilation have been shown in the above two examples by taking only one parameter as variable while keeping the other one fixed.

CHAPTER 13

Twin Paradox and Travelling into Future, Perpetrated by Einstein, are Misinterpretations

Nobody has grown up in physics, over the last more than a century, without hearing/teaching or fantasizing about these themes. These are spoken of in the context of theory of Special Relativity, which was set forth by Einstein in 1905.

It goes like this – Suppose there are twin sisters named 'E' and 'S'. When 'S' goes on a spaceship travelling at a speed close to that of light, her aging would be slower (as the time ticks slower for her according to the theory) and when she returns back, she would be visiting earth in future. For instance, if she had travelled for a year, she would be older by only a year whereas her twin sister 'E' would be much older or could even be dead. From the Special Relativity relations for time, one could calculate exactly how fast or for how long one needed to travel in the spaceship to take a leap by certain numbers of years into future.

Even Einstein himself restated and elaborated on this result in 1911 as follows.

"If we placed a living organism in a box ... one could arrange that the organism, after any arbitrary lengthy flight, could be returned to its original spot in a scarcely altered condition, while corresponding organisms which had remained in their original positions had already long since given way to new generations. For the moving organism, the lengthy time of the journey was a mere instant, provided the motion took place with approximately the speed of light."

Now, an obvious question arises as to why 'E' would turn older to 'S' because of latter's travel. Guided by common experience and the principle of relativity, the twin 'S' may consider herself to be stationary and her sister 'E' as moving with the same speed in opposite direction. All would have to agree to her. Under such a situation,

'S' should turn older to 'E' by the same logic. This is paradoxical as the same twin cannot be older as well as younger to the other one. This is known as Twin Paradox.

Since advent of special relativity, explanations have been attempted innumerable times by experts all over the world without giving a satisfactory answer. Even Einstein's statement quoted above is incorrect interpretation of his own theory, which supports the paradox rather than solving it.

This is a burning example of how twists were imparted to the Special Relativity relations that lead to incorrect conclusions. Let us now analyse it correctly.

When one recalls chapter 12 (Behaviour of Measuring-rods and Clocks in Motion) of Einstein's book, it becomes amply clear that the observers on earth would record, if they could, the time on spaceship in a larger proportion (by Lorentz Factor). That is to say a second on the spaceship would be recorded in many seconds on the earth and the magnitude of 'many' would be dependent on the speed of spaceship.

In fact, just the reverse would be happening when observers on the spaceship were recording time on the earth. They would observe that a second on earth passes in many seconds on the spaceship. Should they conclude they are aging faster than their friends on the earth? Certainly not!

It is only a phenomenon of cross observation and here is no effect on the individual rates of passage of time, either on the earth or on the spaceship. When 'S' returns back from the spaceship, she would be as older as 'E'.

To clarify it further, when 'S' compared her clock timings with those of 'E' on her own (E's) clock, both of them would be in agreement. It is only when 'S' compared her **observation of the E's clock** with her own clock or E's own clock, or vice versa, the disagreement emerges.

This is because a time interval on the spaceship, when observed from earth, or vice versa, would pass in a longer period of earth's own time or spaceship's own time, in accordance with the special relativity.

So, there is neither any conflict nor a paradox and the statement
Should be like this - Suppose the twin 'S' goes for a year on a spaceship travelling at a speed close to that of light and having a clock observable by 'E'. When she returns after **a year of hers observed by 'E' (which would be different from her own year or E's own year)**, she will find that both - she as well as her

sister 'E' - are older by more than a year, as she and 'E' would have to wait longer for completion of **such a year**. How much longer will depend on speed.

Twins with Equal Height Paradox:

The discussion so far has dealt with only the time part. However, to cover the phenomena in totality, the other dimension i.e. length should also have been added by stating 'Twins with Equal Height Paradox'.

In the aforesaid example, suppose 'E' and 'S' had equal heights too and the spaceship was moving along the height of 'S'. Further, both the sisters have identical electronic height measuring instruments which are not affected by speed, gravity etc. and can remotely measure the height of the other sister by say, quantum entanglement.

According to the same interpretation, in respect of length, as for the time (age), each of the two sisters - 'E' as well as 'S' - would find that the other sister has gone shorter too, in addition to getting younger, as the relative motion applied in the same way to both. This leads to the expanded paradox the section talks about.

The correct interpretation is similar to the one for the age presented above. Both are phenomena of cross measurement and there would be no change, either in their heights or in their age.

CHAPTER 14

Relativity Can Neither Extend Muon's Lifespan Nor Shorten its Path

Muons, the negatively charged unstable elementary particles, generated from collision of cosmic radiation with the Earth's atmosphere, have a mean lifespan of 2.2 microseconds. Their generation is estimated to be at a distance of 15-100 km up in the atmosphere, after considering may rounds of cascading collisions descending downwards.

1. The Question:

The question is: even a muon with a speed as high as $0.99999c$ can travel only a distance of $0.99999 \times 300000 \times 2.2 \times 10^{-6} = 0.660$ km = 660 m before decaying. With this magnitude of travel capacity, even those generated at the heights of 15 km, are detected on the Earth's surface. Such phenomena cannot be explained by the classical kinematics/dynamics.

However, Relativity has been brought in to explain it, and it has gained stronghold over the last 87 years since discovery of the particle.

2. Current Incorrect Explanations:

Currently, there are two approaches to offer the explanation.

2A. Time Dilation Approach:

Muons live for 2.2 microseconds of their time, which observers on the Earth would measure $2.2/\sqrt{1 - v^2/c^2}$ microseconds. With the figures assumed above, this works out to 492 microseconds, which translates into a travel distance of $0.99999 \times$

$300000 \times 492 \times 10^{-6} = 147.6$ km. This is more than even the highest estimates of 100 km. Thus it makes arrival possible on Earth, of almost all the muons generated. A lower assumed speed of muon would result into a lower distance of travel but the theme remains the same.

2B. Length Contraction Approach:

For the muon, in its own frame, the entire length of Earth's atmosphere, say 100 km, is treated as moving upwards, and therefore, it appears to it as contracted by the factor i.e. $\sqrt{1 - v^2/c^2}$. This leads to a contracted length of only 0.447 km, which is easily covered by the muon within its lifespan of 2.2 microseconds.

The above explanations come from, either a lack of understanding of, or a motivated departure from, the theory in its correct form.

3. Correct Interpretation Leaves the Problem Unresolved:

Even though the differing measurements of a segment of time or distance, done simultaneously from the two different frames - of Earth and muon itself – could well be reconciled by Relativity, yet nothing could be done (from what Relativity has to offer) to increase the lifespan of the muons, or shorten their paths.

It must be borne in mind that the **muon is nothing more or less than the origin of the moving frame**, with Earth treated as the stationary frame, in the context of the Lorentz transformation that forms the basis of all kinds of explanations.

Keeping the setup of derivation of the Lorentz transformation in mind, the correct interpretations from both the approaches are as follows.

3A. Time Dilation Approach:

The lifespan of 2.2 microseconds has been worked out in labs on the Earth in say, stationary frame conditions. Therefore, even in muon's own frame of reference, which is stationary, the same time (and lifespan) holds.

Suppose, another muon is also created in a lab on the Earth, just below and simultaneous with the one created in the atmosphere.

When the lab muon is considered stationary, the atmospheric muon is to be considered moving towards it with a speed of $0.99999c$. Similarly, when the atmospheric muon is considered stationary, the lab muon is to be considered as moving towards the former with the same speed i.e. $0.99999c$. Whichever way one considers, the fact remains that the relative velocity of the moving frame is $0.99999c$.

Recalling the derivation of Lorentz transformation (which is used to argue out the time dilation, leading to increased lifespan of muons), the distance travelled by the moving frame is to be worked out by multiplying the relative velocity with the time of the frame considered stationary, and not with the time of the moving frame, as recorded by the stationary frame, which is incorrectly being done at present.

No matter which muon is considered stationary, only the time of the stationary frame, which is 2.2 microseconds in either case, is to be taken to work out the distance travelled by the moving frame (muon). This distance works out to be only 660 m.

When the moving frame/clock (muon) itself ceases to exist after 2.2 microseconds of Earth's time, and also of the muons' own time, how can these travel for 15-100 km? Thus the problem still remains, and Relativity cannot answer it.

Unfortunately, the current interpretation is similar to the misinterpreted phenomena like 'twin paradox' and 'travelling in to future'. Einstein's understanding of his own theory (of Special Relativity and time dilation) was so limited that he propounded these dream selling phenomena as its outcomes.

According to the correct interpretation of the same Relativity, time dilation or slow aging appear to be happening only due to cross-measurements (parameters of one frame getting measured by the other frame), and these do not occur in the observer's own frame. The changed measurements of the other frame parameters do not affect the aging rate or lifespan of that (the other) frame.

In the instant example, though each of the two muons would find the other one's lifespan to have increased multi fold by a factor of (492/2.2), yet their own lifespans would be in their own time, and not in others' time. Their own time would continue to be 2.2 microseconds, and each of them would decay much before this misplaced privilege of living on others' time plays out fully for them. At the time of its own decay, each one would record the other one's time as only (2.2/492) microseconds due to time dilation, but this finding would not allow anyone to live any longer.

3B. Length Contraction Approach:

The Lorentz transformation is used to calculate the "contracted" length of Earth's atmosphere, from the muon's frame of reference. It may be recalled that the contraction by a factor of $\sqrt{1 - v^2/c^2}$ would occur only when the moving muon takes, in its frame, the same time at both the ends of the atmosphere (i.e. in a snapshot). The muon would see the same contracted length of the atmosphere, no matter where it was during course of its travel, as the difference of time at the two ends was taken as zero in the muon's frame.

However, the fact which is missed in such an interpretation is as follows.

For the muon to get the same time, in its frame, at both the ends of atmosphere, the corresponding time difference in the Earth's frame should be vx/c^2 according to the same Lorentz transformation. This works out to $0.99999 \times 100/300000$ sec = 333 microseconds in Earth's time. The moving frame i.e. muon has to be running for such a magnitude of time (in Earth's time) to record the time difference as zero. This obviously is not possible in view of a much shorter lifespan of 2.2 microseconds for the moving frame i.e. muon. The muon would not live long enough to see this contracted length of atmosphere.

4. Options Available:

With due regards to all concerned, I have no hesitation in saying that we have been fooling ourselves on time dilation for more than a century now. Time dilation, if at all it existed in the moving observer's own frame, the theory of Relativity, in its present form, is unable to explain it. As a result, we have to look for other theories, or an upgraded version of Relativity, to explain it. The correction could also lead to fixing of some missing links in particle physics.

Out of the known principles, an explanation from Quantum Mechanics could well be attempted. A quantum phenomena similar to that leading to fusion of similarly charged, speeding protons in Sun could well be playing out here. Earth, being a large neutral reservoir, could cause collapse of wave functions of electrically charged muons, carrying the right magnitude of energy/momentum, in such a way that it resulted into such large lifespans, or their swift appearance on the Earth.

SUGGESTIONS

CHAPTER 15

The Set of Postulates is Required to be Expanded

Oblique Observation Setup:

The second postulate of Special Relativity states as follows-

The speed of light in free space has the same value c in all inertial frames of reference.

An inertial frame is defined as a frame of reference which is at rest or is moving with a uniform velocity along a straight line. That means the direction of motion of such a frame does not disqualify it from being an inertial frame.

Resultantly, even an inertial frame moving in a direction that is oblique to that of the light signal should find the speed of light c as the same in the direction of its observation i.e. from the origin of the frame to the light signal. This has also been verified by experiments.

Keeping this in mind, the Lorentz Transformation can be derived for oblique observations i.e. when the direction of observation is oblique to that of the moving light signal. This has been done in chapter 16.

The transformation relations so arrived at do conform to Maxwell's EM wave equation, which is a strong validation for them. The exercise is presented in chapter 17.

With such a scheme that permits oblique observations, the inconsistency in respect of events falling outside the line of observer/frame's motion discussed in Observations 8 of chapter 4, disappears.

Further, with this scheme, it is observed that the Wigner-Thomas Rotation disappears. The exercise has been presented in chapter 18.

Thus the inertial frames moving obliquely to the light signal under observation, resulting into oblique observations, fully conform to the 2nd postulate and also, remove some inconsistencies. So, why should such fames be kept out of consideration in special relativity?

Some argue that Thomas precession, a relativistic correction, is required to explain the spin of elementary particles and therefore, its disappearance in the oblique observation setup is counter-productive. However, the flip side is - calculation of Thomas precession for elementary particles, using the Lorentz transformation that is valid only for light, is itself incorrect. Thus the argument falls apart.

Further, continuance of the current scheme in respect of the events falling outside the line of motion of the observer results into a serious inconsistency.

So there is a need to discuss the pros and cons of the two setups and arrive at the solution. If it is decided to reject the oblique observation setup and continue with the current setup, the following would require to be added to the set of postulates-

1. The transformations will be governed by the distance of event along the line of motion of the moving frame/observer.
2. The transformation factor will be the same as that for co-directionally moving frames.

Application of Lorentz Transformation to Material Bodies:

The Lorentz Transformation have not been derived for events other than those of light. However, we continue using them extensively for all events. As a result of this, the Lorentz Factor γ can take any value without affecting the transformation relations for such events. Despite this, its expression $1/\sqrt{1 - v^2/c^2}$ that is valid only for light is also adopted for other events. The chapters 5, 6 and 7 amply highlight the implications of such an action.

If it is still decided to retain the same Lorentz Factor for transformation of all events, certain validations required, which are given in chapter 9. Even, after successful validations, the following would still be required to be added to regularise the current practice.

1. The ratio of transformation of distance and time emerging out at any given velocity of the moving frame, to maintain the constancy of light speed in vacuum c, will hold good for all distance- time sets.

The Solution:

Three additions have been identified above for augmentation of the existing set of two postulates to validate the current practices in Special Relativity, subject of course to the validation required, as discussed in chapter 9 (for sr. no.3 below).

Further, it is clear that the three additions can be merged in to two. Thus the total set of postulates should be expanded as follows.

1. The laws of physics are the same in all inertial frames.
2. The speed of light in free space has the same value c in all inertial frames of reference.
3. The transformations will be governed by the distance of event along the line of motion of the moving frame/observer.
4. The ratio of transformation of distance and time so emerging out at any given velocity of the moving frame, to maintain the constancy of light speed in vacuum c, will hold good for all distance- time sets.

CHAPTER 16

Lorentz Transformation with Oblique Observation of a Moving Light Signal

Abstract:

Lorentz transformation for distance and time, as we see today, have emerged out of, and therefore, are valid only for parallel observation of moving light (signal) i.e. when the observer (moving frame) moves in the same direction as that of the light signal). Surprisingly, no need for such relations has ever been felt for situations where the event (distance-time set) of light lies outside of the line of motion of the observer or vice versa. The probable reason is that the parallel component of either the event location or the frame velocity is taken and the existing relations applied. Einstein institutionalized the former (parallel component of event location) in his book. This approach is, however, contradictory to the very postulate of constancy of light speed in all inertial frames. This article shows how and also offers the solution by presenting a new derivation which expectedly leads to different relations.

Misplaced Complacency on Parallel Observations:

The complacency pervades everywhere because there is a widespread belief that (1) such situations would hardly arise and if at all they did, (2) a simple step of taking the parallel component of the event's location would enable one to utilise the existing relations. The solution has been institutionalized by none other than Einstein himself by his 1916 book "Relativity: The Special and The General Theory", wherein the derivation of Lorentz transformation (for parallel observation) has been presented in Appendix I.

After working out the relations at eq.8, he has mentioned as to what is to be done for events falling outside the *x*-axis (the line of motion of the light signal as well as of the moving frame).

The **relevant piece of text is reproduced below in italics**, followed by my observations in normal font, between two dotted lines.......

The equations (6) and (7b) determine the constants a and b. By inserting the values of these constants in (5), we obtain the first and the fourth of the equations given in Section 11.

$$\left.\begin{array}{l} x' = \dfrac{x-vt}{\sqrt{1-\dfrac{v^2}{c^2}}} \\ t' = \dfrac{t-\dfrac{v}{c^2}x}{\sqrt{1-\dfrac{v^2}{c^2}}} \end{array}\right\} \dots (8)$$

Thus we have obtained the Lorentz transformation for events on the x-axis. It satisfies the condition

$$x'^2 - c^2 t'^2 = x^2 - c^2 t^2 \dots (8a)$$

The extension of this result, to include events which take place outside the x- axis, is obtained by retaining equations (8) and supplementing them by the relations

$$\left.\begin{array}{l} y' = y \\ z' = z \end{array}\right\} \dots (9)$$

In this way we satisfy the postulate of the constancy of the velocity of light in vacuo for rays of light of arbitrary direction, both for the system K and for the system K'. This may be shown in the following manner.

We suppose a light-signal sent out from the origin of K at the time t = 0. It will be propagated according to the equation

$$r = \sqrt{x^2 + y^2 + z^2} = ct$$

or, if we square this equation, according to the equation

$$x^2 + y^2 + z^2 - c^2t^2 = 0 \ldots \ldots (10)$$

It is required by the law of propagation of light, in conjunction with the postulate of relativity, that the transmission of the signal in question should take place — as judged from K1 — in accordance with the corresponding formula

$$r' = ct'$$

Observations:

Events outside the **x-axis** would be generated when the light signal, taken in his derivation, moved in a direction oblique to **x-axis.**

Under such a setup, the statement of eq.(9), with only *x* taken for transformation, would be true and in conformity with the Lorentz Transformation Condition, only when the postulate was modified to the following.

"The component of speed of light in vacuo **along the direction of motion of the inertial frame** remains constant (as **c**)".

However, if it was done, it would be wrong for reasons explained below.

Consider the situation faced by the stationary observer. He/she takes measurements of speed (of the light signal) in the two directions. He/she would definitely measure speed as c in the direction of propagation of light. Concurrently, however, **for application of the transformation relations in accordance with their derivation,** the speed in the direction of motion of the other observer (moving along x-axis) has also to be the same i.e. c.

This creates an impossible situation, as a speed in any arbitrary direction and its component in some other direction are being attempted to be made equal. If the event causing light signal was assumed to be casting a shadow on x-axis, could the speed of light signal and that of its shadow be the same? Obviously not!

In other words, in the stationary frame, the light signal which actually travels a distance of $\sqrt{x^2 + y^2 + z^2}$ in a time t would be considered to travel by only x with the same speed c for transformation of the entire distance and time. It prima facie sounds inappropriate.

It may be noted that this is the case of a **single** ray/signal of light whose speed is measured in two different directions, and these are bound to be different. It is different from speeds of **two** different rays/signals of light measured in their own direction of motion, which would always be the same as c.

In the instant case, since the event outside the x-axis represents one signal of light moving in a particular direction, it cannot be assigned the same speed in two different directions.

Thus application of the relativity relations on the x-component of the event's location is in violation of the very assumptions (of derivation), as the speed of the moving signal in x-direction is not c.

On the other hand, if the postulate is to be maintained in its present form, statement of eq.(9) is incorrect, as the directions of measurement of signal's distance from the origins of the two frames (which undergo transformation along with their respective times) are different.

With the present postulate, it is the total distance **r** and not one of its components **x** which would undergo transformation along with tits corresponding time **t**, so as to maintain the constant speed **c**.

Therefore, a clear choice needs to be made whether the cardinal postulate of the theory is to be maintained or surrendered in favour of the Lorentz Transformation Condition for parallel observations .I consider the former to be far more important than the latter.

The Derivation for Out-of-Line Events (Oblique Observation):

The derivation of relations has been done **with the same postulate** i.e. constancy of light speed in vacuum in all inertial frames.

However, it has been made more robust by application of the essence of relativity i.e. reciprocity, leaving no room for objections.

The relations emerging out would be very different from what our eyes are tuned to see, having quite a good load of trigonometric functions.

However, as you would see, one gets the same relations as existing today, on substitution of the angle(s) with 0 or π for parallel observations.

Let us now start with the derivation.

Refer the following fig.

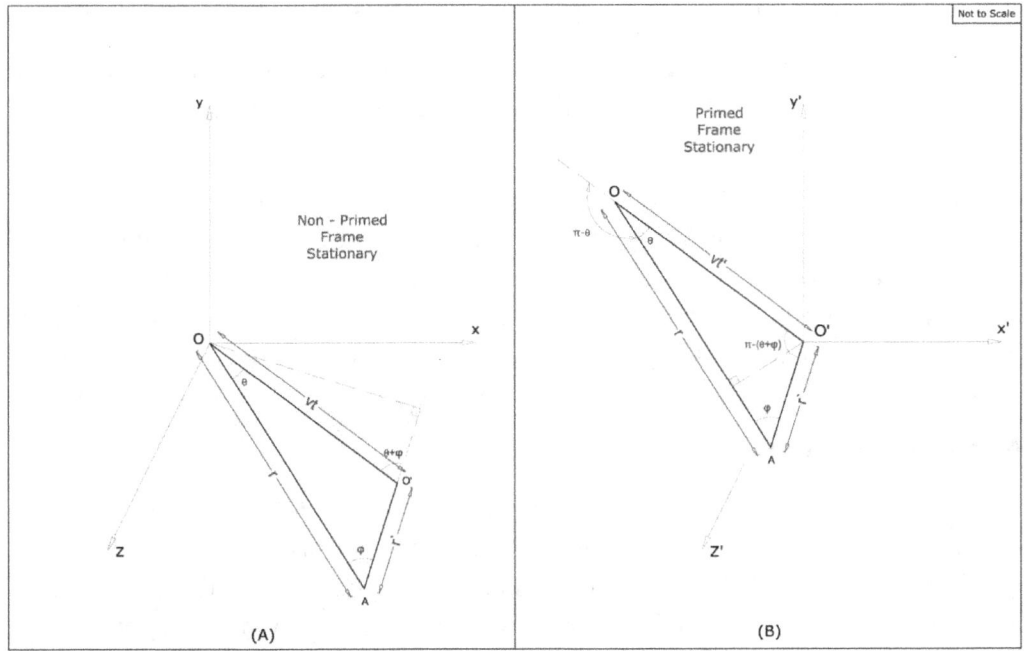

Assumed Setup for Derivation:

The following setup and assumptions are made in a way similar to the original derivation but with a more general case of three dimensional event location as well as frame velocity instead of a unidirectional setup.

1. The corresponding three axes of frames K and K' are permanently aligned with one another and at the start, their origins are coincident and concurrent i.e. $x = 0, y = 0, z = 0, t = 0$ and also $x' = 0, y' = 0, z' = 0, t' = 0$.

2. The distance from origin and time of an event in three dimensional space in frame K are designated as r and t respectively and those in frame K' for the same event are designated r' and t'.

3. The frame of observation K' is moving with respect to the frame K in three dimensional space with a uniform relative velocity v on the side of the event i.e. the angle between $\vec{r}$ and $\vec{v}$ is acute.

4. The angle between $\vec{r}$ and $\vec{v}$ is θ and that between $\vec{r'}$ and $\vec{r}$ is φ. Consequently, the angle between $\vec{r'}$ and $\vec{v}$ becomes $(\theta + \varphi)$.

5. All the events (distance-time sets) are reckoned from the start defined at sr. no. (i) above. Consequently, the light signal is also assumed to start its motion from the same origin.

The light signal is at point 'A' at the given time in the above fig. The event distance triangle OAO' along with the subtended angles is shown in Fig. (A) and Fig. (B) considering respectively the non-primed and the primed frame as stationary, to work out the distance and time in the other frame which is considered moving. The origins and axes of the non-primed and the primed frames are O-X-Y-Z and O'-X'-Y'-Z' respectively.

Notation U taken in place of c:

I consider that the speed of light in vacuum is only a visible, measurable form of Nature's Universal Communication Speed, termed as U, which pervades everywhere perpetually as manifested in speed of transmission of electromagnetic waves, actions, forces and fields etc. Motion of light, though relevant here, does not exist when we consider events other than those of light in other concepts, such as in spacetime.

Further, the speed is not limited to just light but is universal. For this reason, I consider U to be more appropriate than c. This has further been elaborated in my book, from where the derivation is taken.

So, the constant U may be treated the same as constant c only.

The Derivation:

From the event distance triangle OAO' of Fig. (A), we may write the relation $r' = r \cos \varphi - vt \cos(\theta + \varphi)$ based on classical kinematics. However, this is not true from the concept of relativity, as the distance and time measured in the moving frame change in such a way as to maintain constancy of speed of light (or Nature's communication). So, let us assume that it will be in some ratio as follows.

$$r' = a(r \cos \varphi - vt \cos(\theta + \varphi))$$

Where a is a constant.

By reciprocity of relativity, the relation may as well be stated the reverse way as follows by referring to Fig. (B).

$$r = a(r' \cos \varphi + vt' \cos \theta)$$

It may be noted that in the second relation, $(\pi - \theta)$ takes the place of $(\theta + \varphi)$ existing in the first equation and similarly in the first relation, $[\pi - (\theta + \varphi)]$ takes the place of θ existing in the second relation.

Now, the above relations for the two frames shall be written together and further derivation shall proceed by performing the same action on both the equations so as to maintain reciprocity of relativity.

$$\left. \begin{array}{l} r' = a(r \cos \varphi - vt \cos(\theta + \varphi)) \\ r = a(r' \cos \varphi + vt' \cos \theta) \end{array} \right\} \quad \ldots \ldots \text{Eq.(1)}$$

From the above two equations, t' may be expressed in terms of t, r and v; similarly, t may be expressed in terms of t', r' and v. By doing so, we get the following relations.

$$\left. \begin{array}{l} t' = a \cos \varphi \dfrac{\cos(\theta+\varphi)}{\cos \theta} \left[t - \left(1 - \dfrac{1}{(a \cos \varphi)^2}\right) \dfrac{r \cos \varphi}{v \cos(\theta+\varphi)} \right] \\ t = a \cos \varphi \dfrac{\cos \theta}{\cos(\theta+\varphi)} \left[t' + \left(1 - \dfrac{1}{(a \cos \varphi)^2}\right) \dfrac{r' \cos \varphi}{v \cos \theta} \right] \end{array} \right\} \quad \ldots \ldots \text{Eq.(2)}$$

The postulate stipulates as follows.

$$\left. \begin{array}{l} r' = Ut' \\ r = Ut \end{array} \right\}$$

On substituting the expressions of t' and t from Eq.(2) and those of r' and r from Eq.(1) in the above relations and then replacing r' with Ut' and r with Ut, we get

$$\left. \begin{array}{l} Ut\cos\varphi - vt\cos(\theta+\varphi) = U\cos\varphi \dfrac{\cos(\theta+\varphi)}{\cos\theta}\left[t - \left(1 - \dfrac{1}{(a\cos\varphi)^2}\right)\dfrac{Ut\cos\varphi}{v\cos(\theta+\varphi)}\right] \\ Ut'\cos\varphi + vt'\cos\theta = U\cos\varphi \dfrac{\cos\theta}{\cos(\theta+\varphi)}\left[t' + \left(1 - \dfrac{1}{(a\cos\varphi)^2}\right)\dfrac{Ut'\cos\varphi}{v\cos\theta}\right] \end{array} \right\}$$

On dividing both the sides of **either** the first equations by $Ut\cos(\theta+\varphi)$ or the second one by $Ut'\cos\theta$, we get the same resultant as follows, thus maintaining the reciprocity of relative motion we began with.

$$\left. \begin{array}{l} \dfrac{\cos\varphi}{\cos(\theta+\varphi)} - \dfrac{v}{U} = \dfrac{\cos\varphi}{\cos\theta} - \left(1 - \dfrac{1}{(a\cos\varphi)^2}\right)\dfrac{U(\cos\varphi)^2}{v\cos\theta\cos(\theta+\varphi)} \\ \dfrac{\cos\varphi}{\cos\theta} + \dfrac{v}{U} = \dfrac{\cos\varphi}{\cos(\theta+\varphi)} + \left(1 - \dfrac{1}{(a\cos\varphi)^2}\right)\dfrac{U(\cos\varphi)^2}{v\cos\theta\cos(\theta+\varphi)} \end{array} \right\}$$

This, on rearranging, results into the following

$$1 - \dfrac{1}{(a\cos\varphi)^2} = \dfrac{v}{U\cos\varphi}\left[\dfrac{v\cos(\theta+\varphi)\cos\theta}{U\cos\varphi} - \cos\theta + \cos(\theta+\varphi)\right]$$

Or,

$$a = \dfrac{1}{\cos\varphi\sqrt{1 - \dfrac{v}{U\cos\varphi}\left[\dfrac{v\cos(\theta+\varphi)\cos\theta}{U\cos\varphi} - \cos\theta + \cos(\theta+\varphi)\right]}}$$

With the value of a now known as above, the values of r or r' can be determined from Eq.(1). Also, with substitution of the above expression of $\left(1 - \dfrac{1}{(a\cos\varphi)^2}\right)$ in eq.(2) above, the transformation relations for t' and t get simplified and the values of t or t' become determinable.

The equations Eq.(1) and simplified Eq.(2) are reproduced below.

$$\left. \begin{array}{l} r' = a(r\cos\varphi - vt\cos(\theta+\varphi)) \\ r = a(r'\cos\varphi + vt'\cos\theta) \end{array} \right\}$$

$$t' = a\cos\varphi \frac{\cos(\theta+\varphi)}{\cos\theta}\left[t - \left(\frac{v\cos\theta}{U\cos\varphi} - \frac{\cos\theta}{\cos(\theta+\varphi)} + 1\right)\frac{r}{U}\right]$$

$$t = a\cos\varphi \frac{\cos\theta}{\cos(\theta+\varphi)}\left[t' + \left(\frac{v\cos(\theta+\varphi)}{U\cos\varphi} + \frac{\cos(\theta+\varphi)}{\cos\theta} - 1\right)\frac{r'}{U}\right]$$

The above set of equations reduce to those for the unidirectional case of the original derivation on substitution of the angle of frame motion to zero i.e. $\theta = 0$ and resultantly, $\varphi = 0$.

Now, for light events, we may use the relations $r = Ut$ and $r' = Ut'$ to further simplify the relations as follows.

(It may be noted that expressions of r' and r are shown along with those of t' and t respectively to show that transformations of distance and time individually take place by the same factor in the given frame.)

$$r' = a\cos(\theta+\varphi)\left[\frac{\cos\varphi}{\cos(\theta+\varphi)} - \frac{v}{U}\right]r$$

$$t' = a\cos(\theta+\varphi)\left[\frac{\cos\varphi}{\cos(\theta+\varphi)} - \frac{v}{U}\right]t$$

And

$$r = a\cos\theta\left[\frac{\cos\varphi}{\cos\theta} + \frac{v}{U}\right]r'$$

$$t = a\cos\theta\left[\frac{\cos\varphi}{\cos\theta} + \frac{v}{U}\right]t'$$

Now, let the transformation factor for transformation from r and t to r' and t' respectively be denoted as λ_1. Resultantly, the transformation factor for the reverse transformation would be $1/\lambda_1$. So, from the above set of relations, we have

$$\lambda_1 = a\cos(\theta+\varphi)\left[\frac{\cos\varphi}{\cos(\theta+\varphi)} - \frac{v}{U}\right]$$

$$\frac{1}{\lambda_1} = a\cos\theta\left[\frac{\cos\varphi}{\cos\theta} + \frac{v}{U}\right]$$

On multiplying separately the LHS and the RHS of the above equations and equating them, we have

$$1 = (a\cos\varphi)^2 \left[1 - \frac{\cos(\theta+\varphi)}{\cos\varphi}\frac{v}{U} + \frac{\cos\theta}{\cos\varphi}\frac{v}{U} - \left(\frac{v}{U}\right)^2 \frac{\cos(\theta+\varphi)\cos\theta}{(\cos\varphi)^2} \right]$$

Or,

$$1 - \frac{1}{(a\cos\varphi)^2} = \frac{v}{U\cos\varphi}\left[\frac{v\cos(\theta+\varphi)\cos\theta}{U\cos\varphi} - \cos\theta + \cos(\theta+\varphi) \right]$$

The above expression is the **same as obtained a few steps earlier**. Thus the reciprocity of transformation is once again established.

The Lorentz Transformation Condition:

The condition in its present form does not hold good for such cases of oblique observation, and is valid only when the event falls on the line of observer's motion.

The condition applicable to such cases has been worked out in my book. However, it is skipped here, being lengthy.

Conformity to Maxwell's Wave Equation:

As light is an electromagnetic wave, it must conform to Maxwell's wave equation. The cardinal term of Special Relativity i.e. Lorentz Factor (for parallel observations, of course) was discovered to strike conformity to the Maxwell's equations from both the frames. As the invariance of U (or c) has been maintained for oblique observations too, the results obtained here should also conform to the equations. This is **found to be true**, and the working is presented in chapter 17. **Consequently, it also re-validates the correctness of transformation relations derived above.**

Conclusion:

The derivation presented above is straightforward, emanating from the most essential attribute of relativity i.e. reciprocity, and therefore, is free from any objections. Further, it offers more general solutions, encompassing even the out-of-line events.

As regards the transformation of out-of-line events, sound theoretical reasons have been presented to establish that such events have to be treated differently to maintain conformity with the current postulate.

It also calls for reworking of the existing relations and interpretations, and also reorientation of our experimental efforts. For example, working out of the Relativistic Doppler's effect, in case of the receiver moving in any arbitrary direction, still uses the same Lorentz Factor γ which does not hold good for such (oblique) observation cases.

Bibliography:

The content is taken entirely from my book "Refining Relativity Part 1 (The Special Theory)".

CHAPTER 17

Maxwell's Equation for EM Waves — Does It Hold Good for Obliquely Moving Observers?

Maxwell's electromagnetic wave equation is derived from his set of four equations for electromagnetic fields which together describe how fluctuations in electromagnetic fields propagate at a constant speed c.

Light is considered an electromagnetic wave and therefore, must follow Maxwell's electromagnetic wave equation.

Since light propagating with a constant speed c is the vehicle for derivation of Lorentz Transformation (also known as Special Relativity relations), its conformity to the wave equation, or otherwise, from both the frames—stationary as well as moving—becomes very important.

While Voigt, Lorentz and Poincare followed the Maxwell's wave equation in their separate contribution to arrive at the present form of Lorentz Transformation, Einstein applied kinematics, though with mistakes, to work it out.

A study of history of Lorentz transformation reveals that certain terms had to be added to the Galilean Transformation relations to arrive at Lorentz Transformation (for parallel observations, of course) so as to strike compatibility with the Maxwell's wave equation from both the frames—stationary as well as moving. For an electromagnetic wave propagating in x-direction under observation from a **frame moving in x-direction itself**, Voigt first suggested in 1897 a correction of vx/c^2 to be applied to the time of stationary frame t to get the corresponding time of the moving frame t' for maintaining compatibility with the wave equation. This was adopted by Lorentz in his 1904 paper and Poincare suggested in 1906 multiplication by a constant factor, which is known as Lorentz Factor γ today.

Attention is drawn to the fact that all the above steps were taken in the context of an observer moving in the **same direction as that of the light wave.** Consequently, the resultant Lorentz Transformation relations, which continue even today without any modification, are valid for parallel observations i.e. when the observer is moving parallel to the light.

A question looms large, which has not attracted anybody's attention so far, is—Does the Maxwell's wave equation hold good for observers moving obliquely (at an Inclination) to the direction of light? If it does, what are the consequences?

Before we go further, let me tell you that I have already derived the Lorentz Transformation relations for such observers in my book titled "Refining Relativity Part 1 (The Special Theory)" available on Amazon. The broad details may also be seen on my blog at https://refiningrelativity.blogspot.com.

As the invariance of c has been maintained for oblique observations too in chapter 5 of my book "Refining Relativity Part 1 (The Special Relativity)", the results obtained there should also conform to the wave equation. I am proud to tell you that **it does with the transformed expressions of distance and time**. The same is shown below.

Let me first inform you that I have used the letter U instead of c for reasons described in my book. Further, the constant used here in place of the Lorentz Factor γ is a and it had to be taken to maintain constancy of light speed in both the frames, with no wave equation in picture at all. Similarly, the correction to time t has been derived there to get t′, rather than making any guess or ansatz.

As the exercise is for obliquely moving observers, I have taken the most general case of a 3D distance denoted by r in place of x. Further, introduction of angles becomes inescapable. The 3D angle between the directions of the observer and the light is θ and that between the directions of the light pulse observed from origins of the stationary and the moving frame is φ. In other words, the angle between v and r is θ and angle between r and r′ is φ. As usual, the primed parameters are for the moving frame. The following fig. show the angles.

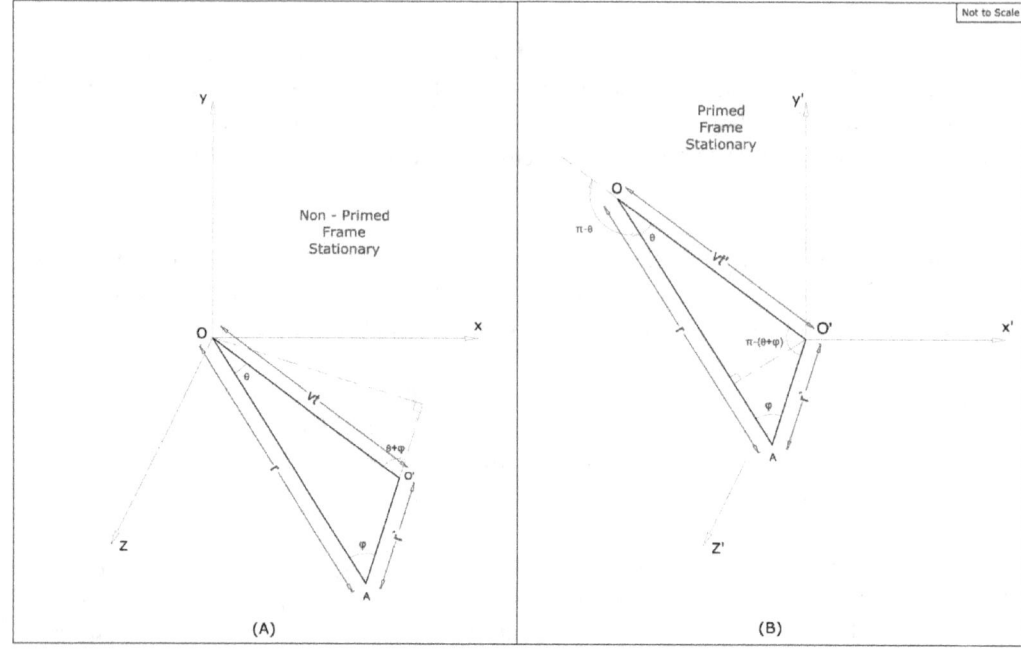

Maxwell's wave equation for the r-direction with Vector Potential $\mathbf{P}$ is as follows.

$$U^2 \frac{\partial^2 \mathbf{P}}{\partial r^2} = \frac{\partial^2 \mathbf{P}}{\partial t^2}. \ldots \text{eq.(1)}$$

From chapter 5 of my book "Refining Relativity Part 1 (The Special Relativity)", we take the following three derived relations

$$\left.\begin{array}{l} r' = a(r \cos\varphi - vt \cos(\theta + \varphi)) \\ t' = a \cos\varphi \dfrac{\cos(\theta + \varphi)}{\cos\theta} \left[t - \left(\dfrac{v \cos\theta}{U \cos\varphi} - \dfrac{\cos\theta}{\cos(\theta + \varphi)} + 1 \right) \dfrac{r}{U} \right] \\ 1 - \dfrac{1}{(a \cos\varphi)^2} = \dfrac{v}{U \cos\varphi} \left[\dfrac{v \cos(\theta + \varphi) \cos\theta}{U \cos\varphi} - \cos\theta + \cos(\theta + \varphi) \right] \end{array}\right\}$$

To simplify working, let us assume constants A and B as follows.

$$A = \frac{\cos(\theta+\varphi)}{\cos\theta} \text{ and } B = \frac{v \cos(\theta+\varphi)}{U \cos\varphi}$$

With these, the above set of equations become as follows.

$$r' = a\cos\varphi\,(r - BUt)$$
$$t' = a\cos\varphi\left[At - \frac{(B-1+A)}{U}r\right]$$
$$\frac{1}{(a\cos\varphi)^2} = 1 - \frac{B^2}{A} + \frac{B}{A} - B = (1-B)\left(1 + \frac{B}{A}\right)$$
......eq.(2)

Now, keeping eq.(1) in mind, we have to work out expressions for $\frac{\partial}{\partial r}$ and $\frac{\partial}{\partial r}$. For this, we will apply the Chain Rule as follows.

$$\frac{\partial}{\partial r} = \frac{\partial}{\partial r'}\frac{\partial r'}{\partial r} + \frac{\partial}{\partial t'}\frac{\partial t'}{\partial r} = a\cos\varphi\left[\frac{\partial}{\partial r'} - \frac{(B-1+A)}{U}\frac{\partial}{\partial t'}\right]$$
$$\frac{\partial}{\partial t} = \frac{\partial}{\partial r'}\frac{\partial r'}{\partial t} + \frac{\partial}{\partial t'}\frac{\partial t'}{\partial t} = a\cos\varphi\left[-BU\frac{\partial}{\partial r'} + A\frac{\partial}{\partial t'}\right]$$
......eq.(3)

On substituting these expressions in eq.(1), we get

$$a^2\cos^2\varphi\left[U^2\frac{\partial^2}{\partial r'^2} - 2U(B-1+A)\frac{\partial^2}{\partial r'\partial t'} + (B-1+A)^2\frac{\partial^2}{\partial t'^2}\right]$$
$$= a^2\cos^2\varphi\left[B^2U^2\frac{\partial^2}{\partial r'^2} - 2ABU\frac{\partial^2}{\partial r'\partial t'} + A^2\frac{\partial^2}{\partial t'^2}\right]$$

On rearranging the terms, we get

$$U^2(1-B)(1+B)a^2\cos^2\varphi\frac{\partial^2}{\partial r'^2} = (1-B)(B-1+2A)a^2\cos^2\varphi\frac{\partial^2}{\partial t'^2} + 2U(1-B)(A-1)a^2\cos^2\varphi\frac{\partial^2}{\partial r'\partial t'}$$

Now, pick up the factor $(A-1)$ of the second term on RHS and add and subtract $\frac{1}{2}\left(-B + \frac{B}{A}\right)$ to it as follows.

$$U^2(1-B)(1+B)a^2\cos^2\varphi\frac{\partial^2}{\partial r'^2}$$
$$= (1-B)(B-1+2A)a^2\cos^2\varphi\frac{\partial^2}{\partial t'^2}$$
$$+ 2U(1-B)\left(A-1-\frac{1}{2}\left(-B+\frac{B}{A}\right)\right.$$
$$\left.+\frac{1}{2}\left(-B+\frac{B}{A}\right)\right)a^2\cos^2\varphi\frac{\partial^2}{\partial r'\partial t'}$$

Take out $\left(-B + \frac{B}{A}\right)$ now from its enveloping bracket as follows.

$$U^2(1-B)(1+B)a^2\cos^2\varphi\frac{\partial^2}{\partial r'^2} = (1-B)(B-1+2A)a^2\cos^2\varphi\frac{\partial^2}{\partial t'^2} +$$
$$U(1-B)\left(2A-2-\left(-B+\frac{B}{A}\right)\right)a^2\cos^2\varphi\frac{\partial^2}{\partial r'\partial t'} +$$
$$U(1-B)\left(-B+\frac{B}{A}\right)a^2\cos^2\varphi\frac{\partial^2}{\partial r'\partial t'}.\ \ldots\ \text{eq.(4)}$$

Let us now take an ansatz that the wave equation of eq.(1) holds good in the primed (obliquely moving) frame also, which will be proved right in a few steps ahead.

So, the wave equation in the primed frame may be written as

$$U^2\frac{\partial^2 P}{\partial r'^2} = \frac{\partial^2 P}{\partial t'^2}$$

Taking square root on both the sides of eq.(1), we have

$$U\frac{\partial}{\partial r'} = \mp\frac{\partial}{\partial t'}$$

Similarly,

$$U\frac{\partial}{\partial r} = \mp\frac{\partial}{\partial t}$$

On substituting the expressions of $\frac{\partial}{\partial r}$ and $\frac{\partial}{\partial r}$ from eq.(3) in the above relation, we get

$$Ua\cos\varphi\left[\frac{\partial}{\partial r'} - \frac{(B-1+A)}{U}\frac{\partial}{\partial t'}\right] = \mp a\cos\varphi\left[-BU\frac{\partial}{\partial r'} + A\frac{\partial}{\partial t'}\right]$$

On rearranging the terms,

$$Ua\cos\varphi\,(1\mp B)\frac{\partial}{\partial r'} = (\mp A + B - 1 + A)\frac{\partial}{\partial t'}$$

The sign $\mp$ may be noted in the above relation, which is valid always for the (−)ve sign and for the (+)ve sign, it is valid only when $(1 + B = 2A + B - 1)$ i.e. when $(A = 1)$. An examination of expression for A will reveal that $(A = 1)$ means parallel observations.

So, we will take the following general relation for the primed frame.

$$U\frac{\partial}{\partial r'} = -\frac{\partial}{\partial t'}$$

Now, using this relation, modify the terms of $\frac{\partial^2}{\partial r' \partial t'}$ in eq.(4) as follows.

$$U^2(1-B)(1+B)a^2\cos^2\varphi \frac{\partial^2}{\partial r'^2}$$

$$= (1-B)(B-1+2A)a^2\cos^2\varphi \frac{\partial^2}{\partial t'^2}$$

$$- (1-B)\left(2A - 2 - \left(-B + \frac{B}{A}\right)\right)a^2\cos^2\varphi \frac{\partial^2}{\partial t'^2}$$

$$- U^2(1-B)\left(-B + \frac{B}{A}\right)a^2\cos^2\varphi \frac{\partial^2}{\partial r'^2}$$

On rearranging the terms,

$$U^2(1-B)\left(1 + B - B + \frac{B}{A}\right)a^2\cos^2\varphi \frac{\partial^2}{\partial r'^2}$$

$$= (1-B)\left(1 + B - B + \frac{B}{A}\right)a^2\cos^2\varphi \frac{\partial^2}{\partial t'^2}$$

From the third relation of eq.(2) we know that

$$\frac{1}{(a\cos\varphi)^2}(1-B)\left(1 + \frac{B}{A}\right) = 1$$

Therefore, the above relation reduces to the following

$$U^2 \frac{\partial^2}{\partial r'^2} = \frac{\partial^2}{\partial t'^2}$$

Which proves the ansatz true.

Thus the Maxwell's wave equation is satisfied in both the frames – stationary as well as moving.

This is also a **reaffirmation** of the transformation relations for distance and time, in respect of light events, derived for **obliquely moving observers.**

CHAPTER 18

Correction of a Flaw in Relativity Eliminates Wigner-Thomas Rotation

Abstract:

Wigner-Thomas rotation arises due to Relativity being founded on only the inertial frames which moved along the line from origin to the event's location, or of light propagation. For observers moving otherwise i.e. in directions oblique to the latter, the setup is converted back to the same (i.e. parallel moving observer), by taking up the distance component parallel to observer's motion for transformation, and ignoring the rest.

The theory does not recognize the inertial frames moving in directions oblique to that of the event, though these too form part of the second postulate of Special Relativity, which is constancy of light speed in all inertial frames.

Two non-collinear boosts together are equivalent to one oblique boost, without compromising the postulate. In other words, two inertial frames moving successively in different directions are together equivalent to one inertial frame moving in the resultant direction, which is oblique to either of the two. Maintaining conformity to the postulate, the resultant relativistic speed can be worked out in this direction. Such a transformation is complete as only a boost, and does not involve any rotation. Further, the transformation of time is worked out with the entire distance covered, and not any of its components.

Introduction:

In Relativity, when two successive non-collinear Lorentz boosts are applied, their resultant is not only a Lorentz boost but also an associated rotation. The rotation obviously emerges again while returning back i.e. when two successive inverse transformations (Lorentz boost) are applied to return to the original state. The total rotation, thus occurring in a pair of onward and inward transformations, is called Wigner-Thomas rotation.

The phenomena go against the very essence of Relativity i.e. reciprocity, which Einstein himself propounded. Einstein's Principle of velocity reciprocity (EPVR) reads as follow.

> *We postulate that the relation between the coordinates of the two systems is linear. Then the inverse transformation is also linear and the complete non-preference of the one or the other system demands that the transformation shall be identical with the original one, except for a change of v to $-v$.*

Contrary to the principle, however, combination of two linear transformations is not linear, and more importantly, one is not able to return back to the original state by inverse transformations, without applying a rotation in addition to the two reverse Lorentz boosts.

Such a disparity arises due to the rule which declares that when the event falls out of line of motion of the observer, only the distance component parallel to the direction of the observer's motion is to be transformed (even if it was zero), **along with the entire time**, leaving untransformed the other (normal) component of the distance.

For example, all the events of an object, moving in a direction perpendicular to that of the observer, are treated as $(0, t)$ for transformation, meaning while the distances along the direction of its motions remain untransformed, the time increases to γt. It leads to a disproportionate reduction in the relativistic velocity of the object, thereby inducing a rotation.

If, however, the two non-collinear velocities are added relativistically with a transformations derived for the resultant oblique direction, the incidence of rotation goes away. This is also a much needed correction in Relativity in order to honour the postulate in its entirety.

The problem, as well as the solution, have been demonstrated with an example in section 1. The example is chosen for addition of two mutually perpendicular velocities, as it simplifies the calculation of Wigner-Thomas Rotation on one hand,

and presents the most adverse setup for the solution proposed, i.e. oblique transformation, on the other.

Further, it has been shown in many chapters that the Lorentz transformations relations are correct only for events of light, the agent chosen originally for derivation, but are only partially correct for others, with the error being proportional to the difference between the distance-time ratio of the event and the speed of light c.

For this reason, the velocities of observers have been chosen very close to that of light in vacuum c in the example taken.

1. The Problem: Wigner-Thomas Rotation:

To keep the exercise simple, let us take up a case of two perpendicular velocities.

The following fig.1 may be referred, which contains three diagrams denoted as (1), (2) and (3).

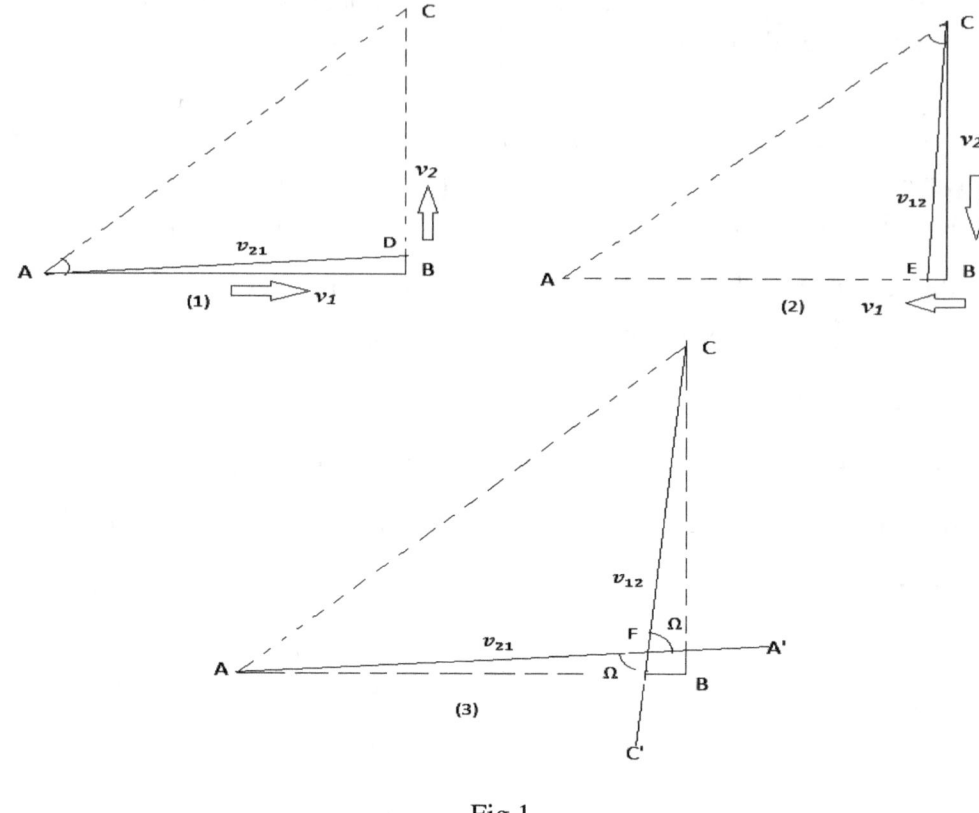

Fig.1

Let there be three observers A, B and C. Observer A is considered stationary. Observer B moves in horizontal direction with a relative velocity v_1 with respect to A, and observer C moves in vertical direction with a relative velocity v_2 with respect to B.

Let us set $c = 1$. So, the velocities v_1 and v_2 are in terms of speed of light in vacuum.

Let v_{21} be the relativistic velocity of C with respect to A. Similarly, let v_{12} be the relativistic velocity of A with respect to C. The two are shown in Fig.1 above.

Einstein's Principle of velocity reciprocity (EPVR) and common intuition demand that $\overrightarrow{v_{21}} = -\overrightarrow{v_{12}}$. However, it is not so, though their magnitudes are equal. Their directions are far apart, as can be seen in Fig.1.

Thus the observer A finds C moving along AD (diagram 1); however, looking inversely, when the observer C looks at A, (s)he finds A moving along AE (diagram 2). As a result, to reconcile with reciprocity, one has to apply a rotation equal to the angle between the two directions, shown as angle Ω in diagram (3). The rotation is called Wigner-Thomas rotation or simply, Wigner Rotation.

The expressions of v_{21}, v_{12} and Ω can obviously be written as follows[1].

$$\overrightarrow{v_{21}} = \overrightarrow{v_1} + \frac{\overrightarrow{v_2}}{\gamma_1} = \overrightarrow{v_1} + \overrightarrow{v_2}\sqrt{1 - v_1^2}$$

$$\overrightarrow{v_{12}} = \overrightarrow{v_2} + \frac{\overrightarrow{v_1}}{\gamma_2} = \overrightarrow{v_2} + \overrightarrow{v_1}\sqrt{1 - v_2^2}$$

$$\|\overrightarrow{v_{21}}\| = \|\overrightarrow{v_{21}}\| = \sqrt{v_1^2 + v_2^2 - v_1^2 v_2^2}$$

$$\sin\Omega = \frac{\|\overrightarrow{v_{21}} \times \overrightarrow{v_{12}}\|}{\|\overrightarrow{v_{21}}\| \|\overrightarrow{v_{12}}\|} = \frac{v_1 v_2 \left(1 - \frac{1}{\gamma_1 \gamma_2}\right)}{v_1^2 + v_2^2 - v_1^2 v_2^2} = \frac{v_1 v_2 \gamma_1 \gamma_2}{1 + \gamma_1 \gamma_2}$$

To see the extent of rotation, let us take velocities v_1 and v_2 very close to c, for reasons mentioned in Introduction.

Let $v_1 = 0.99997$ and $v_2 = 0.99995$.

Therefore, $\gamma_1 = 258.20$ and $\gamma_2 = 200.00$

From the above-mentioned relations, we get

$$v_{21} = v_{12} = 0.999999$$

$$\Omega = 89.19 \text{ degrees}$$

The rotation angle is as high as 89 degrees, in total disregard of the essence of Relativity i.e. reciprocity, **and thus is a roaring alert for a flaw in Relativity.**

2. The Solution: Adoption of Inertial Frames Moving in Oblique Directions:

The frames moving with uniform velocity, but in directions different from that of the light signal under observation, are also inertial frames. Further, such frames too find the speed of light, moving in a particular direction, the same as c while measuring it directly from the frame, in an oblique direction. Thus conformity to the second postulate of Relativity, i.e. constancy of speed of light in all inertial frames, is maintained by such frames.

Ignoring such frames tantamount to deviating from the postulate, which manifests itself in the form of Wigner-Thomas Rotation. On the other hand, working out transformation directly, along the direction from such a frame to the light signal, is free from any rotation.

The derivation is presented below

IMP.: Before proceeding further, it is clarified that the derivation ahead is for transformation of events of light in oblique direction. The same relations would be used on the material frames/objects for addition of velocities, as is being done for the collinear transformation case. To keep the errors occurring on this account to the minimum, frames moving at velocities very close to speed of light have been considered (refer v_1 and v_2 in the previous section).

The following Fig.2 may be referred.

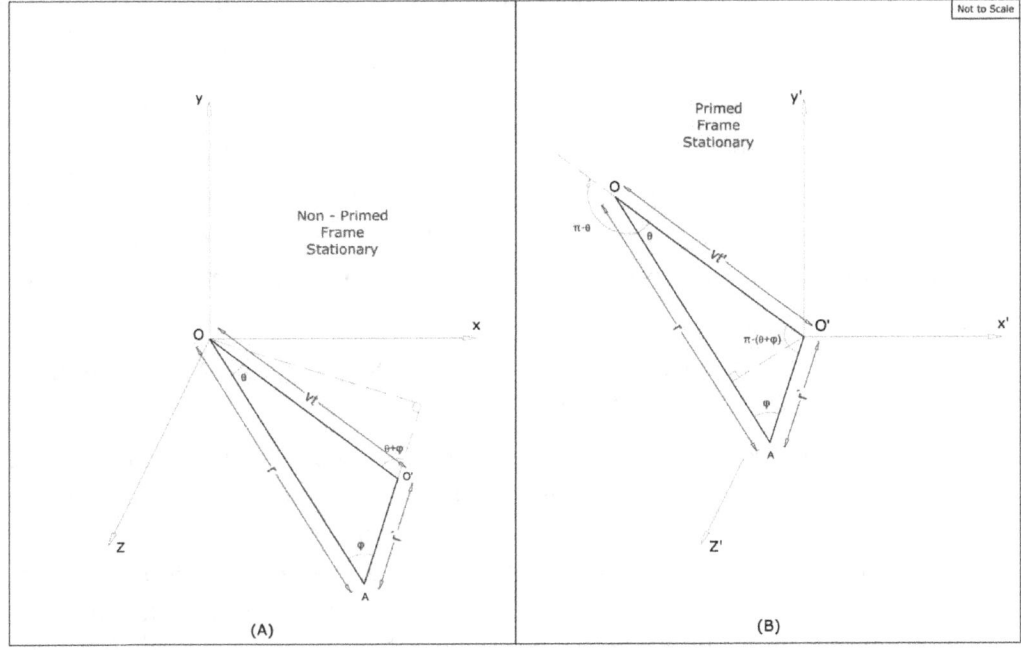

Fig.2

2.1. Assumed Setup for Derivations:

The following setup and assumptions are made in a way similar to the "simplified" derivation presented by Einstein [2], but with a more general case of three dimensional event location as well as frame velocity, instead of a unidirectional setup.

6. The corresponding three axes of frames K and K' are permanently aligned with one another and at the start, their origins are coincident and concurrent i.e. $x = 0$, $y = 0, z = 0, t = 0$ and also $x' = 0, y' = 0, z' = 0, t' = 0$.

7. The distance from origin and time of an event at A in three dimensional space in frame K are designated as r and t respectively and those in frame K' for the same event are designated r' and t'.

8. The frame of observation K' is moving with respect to the frame K in three dimensional space with a uniform relative velocity v on the side of the event i.e. the angle between $\vec{r}$ and $\vec{v}$ is acute.

9. The angle between $\vec{r}$ and $\vec{v}$ is θ and that between $\vec{r'}$ and $\vec{r}$ is φ. Consequently, the angle between $\vec{r'}$ and $\vec{v}$ becomes $(\theta + \varphi)$.

10. All the events (distance-time sets) are reckoned from the start defined at sr. no. (i) above. Consequently, for events generated by moving objects, including light, the object is also assumed to start its motion from the aforesaid start.

11. The event occurs in the stationary non-primed frame K, and therefore, its location is anchored to the origin of this frame.

> **NOTE:** It may be noted that the values of θ and φ, and also the shape of the triangle of Fig.2(B), would be different when the event occurred in the primed frame K'. In such cases, the event location gets anchored to the origin of the primed frame, and for applying the relations derived below, the primed frame K' should be considered as stationary, and the non-primed frame K as moving with a velocity $-v$.

The event distance triangle OAO' along with the subtended angles is shown in Fig.2(A) and Fig.2(B) considering respectively the non-primed and the primed frame as stationary, to work out the distance and time in the other frame which is considered moving. The origins and axes of the non-primed and the primed frames are O-X-Y-Z and O'-X'-Y'-Z' respectively.

2.2. Derivation:

From the triangle OAO' of Fig.2 (A), we may write the relation $r' = r\cos\varphi - vt\cos(\theta + \varphi)$ based on classical kinematics. However, this is not true from the concept of relativity, as the distance and time measured in the moving frame change in such a way as to maintain constancy of speed of light (or Nature's communication). So, let us assume that it will be in some ratio as follows

$$r' = a(r\cos\varphi - vt\cos(\theta + \varphi))$$

Where a is a constant.

By reciprocity of relativity, the relation may as well be stated the reverse way as follows by referring to Fig.2 (B).

$$r = a(r' \cos \varphi + vt' \cos \theta)$$

Now, the above two relations shall be written together and further derivation shall proceed by performing the same action on both the relations so as to maintain reciprocity of relativity.

$$\left. \begin{array}{l} r' = a(r \cos \varphi - vt \cos(\theta + \varphi)) \\ r = a(r' \cos \varphi + vt' \cos \theta) \end{array} \right\} \ldots \ldots (1)$$

From the above two equations, t' may be expressed in terms of t, r and v; similarly, t may be expressed in terms of t', r' and v. By doing so, we get the following relations.

$$\left. \begin{array}{l} t' = a \cos \varphi \, \frac{\cos(\theta+\varphi)}{\cos \theta} \left[t - \left(1 - \frac{1}{(a \cos \varphi)^2}\right) \frac{r \cos \varphi}{v \cos(\theta+\varphi)} \right] \\ t = a \cos \varphi \, \frac{\cos \theta}{\cos(\theta+\varphi)} \left[t' + \left(1 - \frac{1}{(a \cos \varphi)^2}\right) \frac{r' \cos \varphi}{v \cos \theta} \right] \end{array} \right\} \ldots \ldots (2)$$

It may be noted that the angles θ and $(\theta + \varphi)$ in the non-primed frame correspond to angles $(\pi + (\theta + \varphi))$ and $(\pi + \theta)$ respectively in the primed frame, and vice versa.

Further, it is important to note two special cases where one of the relation (1) falls short of the term of time, leading to invalidity of the relations (2). When $\theta = n\pi/2$ or $(\theta + \varphi) = n\pi/2$, RHS of one of the relations (1) loses the term of time – either of t or of t', and this in turn makes the relations (2) invalid, which also gets reflected in infinite values of t' and t respectively. **In such cases**, the results become extreme and inconsistent for events of light as well as all others. So, the workaround is to apply an extremely small correction of, say ± 0.000001, or even lower, to these angle values to keep the relations working.

Going further, the derivation is carried out simultaneously for both the frames to demonstrate that the same results are finally obtained by working in either of the frames. The expressions for the two frames are placed one below the other and enclosed by a curly bracket on right hand side.

The postulate stipulates

$$\left.\begin{array}{l} r' = ct' \\ r = ct \end{array}\right\}$$

On substituting the expressions of t' and t from relations (2) and those of r' and r from relations (1) in the above relations and then replacing r' with ct' and r with ct, we get

$$\left.\begin{array}{l} ct\cos\varphi - vt\cos(\theta+\varphi) = c\cos\varphi\,\dfrac{\cos(\theta+\varphi)}{\cos\theta}\left[t - \left(1 - \dfrac{1}{(a\cos\varphi)^2}\right)\dfrac{ct\cos\varphi}{v\cos(\theta+\varphi)}\right] \\[2mm] ct'\cos\varphi + vt'\cos\theta = c\cos\varphi\,\dfrac{\cos\theta}{\cos(\theta+\varphi)}\left[t' + \left(1 - \dfrac{1}{(a\cos\varphi)^2}\right)\dfrac{ct'\cos\varphi}{v\cos\theta}\right] \end{array}\right\}$$

On dividing both the sides of either the first equations by $ct\cos(\theta+\varphi)$ or the second one by $ct'\cos\theta$, we get the same resultant as follows, thus maintaining the reciprocity of relative motion we began with.

$$\left.\begin{array}{l} \dfrac{\cos\varphi}{\cos(\theta+\varphi)} - \dfrac{v}{c} = \dfrac{\cos\varphi}{\cos\theta} - \left(1 - \dfrac{1}{(a\cos\varphi)^2}\right)\dfrac{c(\cos\varphi)^2}{v\cos\theta\cos(\theta+\varphi)} \\[2mm] \dfrac{\cos\varphi}{\cos\theta} + \dfrac{v}{c} = \dfrac{\cos\varphi}{\cos(\theta+\varphi)} + \left(1 - \dfrac{1}{(a\cos\varphi)^2}\right)\dfrac{c(\cos\varphi)^2}{v\cos\theta\cos(\theta+\varphi)} \end{array}\right\}$$

This, on rearranging, results into the following

$$1 - \dfrac{1}{(a\cos\varphi)^2} = \dfrac{v}{c\cos\varphi}\left[\dfrac{v\cos(\theta+\varphi)\cos\theta}{c\cos\varphi} - \cos\theta + \cos(\theta+\varphi)\right]$$

Or,

$$a = \dfrac{1}{\cos\varphi\sqrt{\left(1 - \dfrac{v\cos(\theta+\varphi)}{c\cos\varphi}\right)\left(1 + \dfrac{v\cos\theta}{c\cos\varphi}\right)}}$$

Having worked out the value of a, the relations (1) and (2) above become the transformation relations for distance and time of an event, from one frame to the other.

It is reiterated the **above relations have come out while maintaining reciprocity of velocity and constancy of speed of light in the directions of OA, as well as of O'A which is oblique to the former.** The transformations change the directions of displacement and velocity, and the time transforms with the entire distance, not any of its components.

Thus no loopholes are left, which may cause a rotation.

2.3. Addition of Velocities:

Moving further, one may work out the relations for addition of velocities also.

It is reiterated that, similar to the co-directionally moving inertial frames in practice today, the same transformation relations emerge out for arbitrary events also, whose distance-time ratio is not equal to c. That, being a different exercise, is not detailed any further here.

Therefore, in working ahead, the ratios r/t and r'/t' are taken as not c but u and u' respectively. To keep the errors occurring on this account to the minimum, however, the velocities considered for addition would be chosen very close to speed of light c, as already done in the previous section 1.

With the known expression of a, we introduce a new constant A for brevity as follows.

$$A = 1 - \frac{1}{(a\cos\varphi)^2} = 1 - \left(1 - \frac{v\cos(\theta+\varphi)}{c\cos\varphi}\right)\left(1 + \frac{v\cos\theta}{c\cos\varphi}\right)$$

On substitution of expressions of r and t from equations (1) and (2), we get

$$u = \frac{r}{t} = \frac{a(r'\cos\varphi + vt'\cos\theta)}{a\cos\varphi \frac{\cos\theta}{\cos(\theta+\varphi)}\left[t' + A\frac{r'\cos\varphi}{v\cos\theta}\right]}$$

On substitution of $r' = u't'$, the expression reduces to

$$u = v\left[\frac{1+\frac{u'\cos\varphi}{v\cos\theta}}{1+A\frac{u'\cos\varphi}{v\cos\theta}}\right]\frac{\cos(\theta+\varphi)}{\cos\varphi} \quad \ldots\ldots (3)$$

The relation (3) is used in the exercise below.

Further, on dividing the two relations of (2), we get

$$\frac{t'}{t} = \frac{a\cos\varphi\,\dfrac{\cos(\theta+\varphi)}{\cos\theta}\left[t - A\dfrac{r\cos\varphi}{v\cos(\theta+\varphi)}\right]}{a\cos\varphi\,\dfrac{\cos\theta}{\cos(\theta+\varphi)}\left[t' + A\dfrac{r'\cos\varphi}{v\cos\theta}\right]}$$

$$= \frac{t}{t'}\left(\frac{\cos(\theta+\varphi)}{\cos\theta}\right)^2 \left[\frac{1 - A\dfrac{u\cos\varphi}{v\cos(\theta+\varphi)}}{1 + A\dfrac{u'\cos\varphi}{v\cos\theta}}\right]$$

Or,

$$t = \left[\frac{\cos\theta}{\cos(\theta+\varphi)}\sqrt{\frac{1 + A\dfrac{u'\cos\varphi}{v\cos\theta}}{1 - A\dfrac{u\cos\varphi}{v\cos(\theta+\varphi)}}}\right] t'$$

On substituting the expression of **u** from (3), the above relation becomes

$$t = \left[\frac{\cos\theta}{\cos(\theta+\varphi)}\sqrt{\frac{1 + A\dfrac{u'\cos\varphi}{v\cos\theta}}{1 - A\left[\dfrac{1 + \dfrac{u'\cos\varphi}{v\cos\theta}}{1 + A\dfrac{u'\cos\varphi}{v\cos\theta}}\right]}}\right] t' \quad \ldots (4)$$

2.4. Reworking of Example of Section 1:

It is pointed out that in all cases of velocity addition below, the setup used in the derivation is strictly followed. Therefore, the first frame/observer, which seeks addition of the velocities of the second and the third frames, in order to know the relative velocity of the third frame with respect to it, is considered stationary and also as the origin.

In other words, the setup of Fig.2 (A) has to be followed, where the stationary observer at O seeks to know the relative velocity of A with respect to it, with the known velocity (u') of A with respect to O', and that (v) of O' with respect to O.

(i) Relative Velocity of C with respect to A, Via B (Fig.1):

In this case, $v = v_1 = 0.99997$, $u' = v_2 = 0.99995$ and $u = v_{21}$ in the direction of AC, whose magnitude is to be found out.

Also, the exterior angle between AB and CB = $(\theta + \varphi) = 90°$. However, to avoid any possible division by zero, we will take this angle as $89.99999°$. Further, with the lengths of AB and CB being in proportion of v_1 and v_2 respectively, we may work out the interior angles by trigonometry, and those would be as follows.

$\angle CAB = \theta = 44.99942°$ and $\angle ACB = \varphi = 45.00057°$

On substitution of these values in the above velocity addition formula (3), one gets

$u = v_{21} = 0.009776163$ in the direction of AC.

The resultant velocity falling to such a level should not surprise, as the corresponding time (at A) has dilated to 144.65541 times of that at B, as per relation (4). Multiplication of the resultant velocity with its time, gives the resultant distance as 1.41417 which matches with the non-relativistic vector addition of velocities i.e. $\sqrt{(0.99997)^2 + (0.99995)^2} = 1.41417$.

(ii) Relative Velocity of A with respect to C, Via B (Fig.1):

In this case, $v = v_2 = 0.99995$, $u' = v_1 = 0.99997$ and $u = v_{12}$ in the direction of CA, whose magnitude is to be found out.

The exterior angle between CB and AB = $(\theta + \varphi) = 90°$. However, maintaining conformity with case (i), we will take this angle also as $89.99999°$. Similar to case (i), the interior angles would be as follows.

$\angle ACB = \theta = 45.00057°$ and $\angle BAC = \varphi = 44.99942°$

As expected, the angle values have got interchanged with each other.

On substitution of these values in the above velocity addition formula (3), one gets

$u = v_{12} = 0.0161881$ in the direction of CA.

As regards its low value comparatively, the reason is similar to case (i) i.e. the corresponding time (at C) has dilated to **87.3585** times of that at B, as per relation (4). Multiplication of the resultant velocity with its time, gives the resultant distance as **1.41417** which matches with the non-relativistic vector addition of velocities i.e. $\sqrt{(0.99997)^2 + (0.99995)^2} = 1.41417$, and also with case (i).

Thus the reciprocity of velocity direction is maintained along AC, and the Wigner-Thomas Rotation goes away. However, the resultant relative velocity is not the same in both the directions, which is inevitable due to different time dilations in frame B, corresponding to different speeds in the two directions.

Conclusion:

The Wigner-Thomas Rotation occurs due to rigidly casting the theory of Relativity to only collinearly moving inertial frames, though the postulate does not impose any such restriction. Such a structure is bound to throw up inconsistencies in non-collinear boosts. This has amply been demonstrated by an example in section 1.

The presented example selects the two non-collinear velocities very close to speed of light in vacuum *c*, for reasons mentioned in the Introduction.

Further, it has been shown in section 2 that if the structure was based on inertial frames moving non-collinearly with the direction of the event, the second postulate i.e. 'constancy of *c* in all inertial frames' was fully complied with, and there was no occurrence of rotation when returning back to the original state after two successive non-collinear boosts onwards and inwards, though the magnitudes were different on account of different magnitudes of time dilation playing out in the two routes.

References:

1. Kane O'Donnel and Matt Visser, "Elementary analysis of the special relativistic combination of velocities, Wigner rotation and Thomas precession", European Journal of Physics 31 (2011) 1033 – 1047, 2011
2. A. Einstein's book "Relativity: The Special and The General Theory)" 1916
3. Author's book "Refining Relativity Part 1 (The Special Relativity)" 2020.
4. W. Engelhardt "On the Origin of Lorentz Transformation", June' 2018

General Relativity

CHAPTER 19

Mistakes of and Questions on General Relativity

1. Spacetime does not conform to the Very Condition it is Built Upon:

The General Relativity bases itself on invariance of lengths of spacetime linear elements, under all frames of observation, including the accelerated ones. The invariant spacetime linear element ds, as defined in equation (1) of his 1916 paper, is as follows.

$$ds^2 = -dX_1^2 - dX_2^2 - dX_3^2 + dX_4^2$$

where X_1, X_2, X_3 are the space coordinates with scales of 1 and X_4 is the time coordinate with a scale of $1/c$ in the local frame considered stationary.

Going back, if a moving frame was moving with a uniform velocity along the space element, the spacetime reduced to a flat one, called Minkowskian spacetime of Special Relativity. Such cases are a very small subset of the general set which the General Relativity claims to address.

Even for the flat spacetime, it permits infinite transformation of distance and time for an event, leading to indeterminacy of a unique transformation. This questions the very concept of spacetime. The detailed exercise on the failures has already been presented in chapter 5.

2. Incorrect and Ad hoc Energy-Mass Conversion and Extrapolation of Gravitational Energy to Total Energy:

Chapter 3 brings out how the energy-mass conversion has been generalized to include all kinds of energy and mass, despite the derivation being mistake-ridden and rudimentary. The relativistic change in wavelength of a photon (like all distances) due to motion of a frame is converted into change in energy of the photon, and this change in energy is treated as the change in kinetic energy (K) which has the form of $\frac{1}{2}mv^2$. Then dividing this change in kinetic energy (ΔK) by $\frac{v^2}{2}$ gives the change in mass i.e. Δm. Thus the change in energy is equated with change in mass. This kind of rudimentary derivation has been treated as the representative of all kinds of mass-energy, including inertial mass and nuclear, chemical energy etc.

Coming to the derivation of General Relativity[1], the following text is quoted in italics from Einstein's derivation (English translated version), which falls between eq. (52) and eq. (53).

The special theory of relativity has led to the conclusion that inert mass is nothing more or less than energy, which finds its complete mathematical expression in a symmetrical tensor of second rank, the energy-tensor. Thus in the general theory of relativity we must introduce a corresponding energy-tensor of matter T_σ^α, which, like the energy-components t_σ [equations (49) and (50)] of the gravitational field, will have mixed character, but will pertain to a symmetrical covariant tensor.*

The system of equation (51) shows how this energy-tensor (corresponding to the density ρ in Poisson's equation) is to be introduced into the field equations of gravitation. For if we consider a complete system (e.g. the solar system), the total mass of the system, and therefore its total gravitating action as well, will depend on the total energy of the system, and therefore on the ponderable energy together with the gravitational energy. This will allow itself to be expressed by introducing into (51), in place of the energy components of the gravitational field alone, the sums $t_\mu^\sigma + T_\mu^\sigma$ of the energy- components of matter and of gravitational field.

The first line of the above quote may be noted. He has concluded. based on his rudimentary derivation quoted above, that even inertial mass is energy. The mass-energy equivalence derivation takes only photons and no inertial mass at all. However, it is extrapolated to cover all mass and going further, a corresponding energy-tensor of matter T_σ^α is also introduced.

The second para of the quote reveals another ad hoc step i.e. substitution of the gravitational field energy with its sum with energy converted from matter, in the field equations (51). It does not require much effort to understand that the gravitational field energy arises out of transformation of distance and time due to observation from an accelerated frame of reference. On the other hand, matter to energy conversion has no basis at all. Therefore, clubbing the two to replace the former in the field equation is another ad hoc step.

Conclusion:

The above facts undermine the theory of General Relativity to a large extent. As a result, the perplexities in the domain of Cosmology, which depends a lot on the GR, can well be understood.

References:

1. A. Einstein "The Foundation of the General Theory of Relativity", Doc 30 of Translated Volume by Princeton University Press, 1997.